Introduction to
NUCLEAR SCIENCE

Introduction to
NUCLEAR
SCIENCE

Jeff C. Bryan

CRC Press
Taylor & Francis Group
Boca Raton   London   New York

CRC Press is an imprint of the
Taylor & Francis Group, an **informa** business

CRC Press
Taylor & Francis Group
6000 Broken Sound Parkway NW, Suite 300
Boca Raton, FL 33487-2742

© 2009 by Taylor & Francis Group, LLC
CRC Press is an imprint of Taylor & Francis Group, an Informa business

No claim to original U.S. Government works
Printed in the United States of America on acid-free paper
10 9 8 7 6 5 4 3 2 1

International Standard Book Number-13: 978-1-4200-6164-2 (Hardcover)

Library of Congress Cataloging-in-Publication Data

Bryan, Jeff C.
    Introduction to nuclear science / Jeff C. Bryan.
        p. cm.
    Includes bibliographical references and index.
    ISBN 978-1-4200-6164-2 (alk. paper)
    1. Nuclear physics. 2. Nuclear chemistry. I. Title.

QC777.B76 2008
539.7--dc22   2008013350

**Visit the Taylor & Francis Web site at
http://www.taylorandfrancis.com**

**and the CRC Press Web site at
http://www.crcpress.com**

Dedication

for
Chris

Contents

Preface

Like many textbooks for specialized classes, this one was born out of frustration with the applicability of existing offerings. This text was initially designed for the nuclear chemistry course I have taught for the past five years. This class is almost entirely populated by nuclear medicine technology majors who need a solid background in fundamental nuclear science, but have not studied math beyond college algebra and statistics. There are many excellent Nuclear Chemistry textbooks available to teach science or engineering seniors or graduate students, but very few for those lacking a more extensive science and math background. Ehmann and Vance's *Radiochemistry and Nuclear Chemistry* served as my primary text for this class for the past four years. It is very well written and nicely descriptive, but it is really intended for radiochemists, not medical professionals. It is also slipping out of date, with no future edition planned. I also learned that some of my students were passing my class without ever cracking the primary text. Clearly, a more appropriate text was necessary.

Writing a nuclear chemistry textbook for nuclear medicine majors seemed too limiting, so this text also includes material relevant to other medical professionals using ionizing radiation for diagnosis and treatment of diseases. In particular, chapters were added discussing radiation therapy and X-ray production. Additionally, it seemed that an introductory nuclear science text should address important contemporary topics such as nuclear power, weapons, and food/mail irradiation. All too often, those speaking passionately on these issues are inadequately versed and resort to emotional appeals while ignoring some of the science. Therefore, this text is also meant to serve as a general primer in all things nuclear for the general public. Very little science and math background is assumed, only some knowledge of algebra and general chemistry.

The more extensive chapters on fission nuclear reactors are also designed to help those that will soon be entering the nuclear workforce. The nuclear power industry in the United States has been static for almost 30 years, but now seems on the verge of renaissance. Nuclear power now appears to be a more environmentally benign source of electricity, and a significant number of new plants may be constructed in the near future. Workers at these plants will need a fundamental understanding of nuclear science as well as a basic idea of how the plant works. This text can serve as a primer for these workers.

Finally, this text can also be of use to scientists making a career start or move to the growing nuclear industry or to the National Laboratory system. I began 13 years of work at Los Alamos and Oak Ridge National Laboratories in 1988. At that point I was thoroughly trained as a synthetic inorganic chemist and was largely ignorant of nuclear science. Some of my motivation in writing this text is to provide scientists in similar situations an easier way to better understand their new surroundings.

The term "nuclear science" is used rather deliberately in the title and throughout the text. It is meant to encompass the physics, mathematics, chemistry, and biology related to nuclear transformations and all forms of ionizing radiation. Fundamentally, nuclear processes are within the realm of physics, but it can also be argued that all of the disciplines mentioned above play an important role in understanding the uses of these processes and the resulting ionizing radiation in today's world.

When this text was initially proposed, some of the reviewers rightly questioned whether my background was appropriate. After all, much of it could be characterized as synthetic inorganic chemistry and crystallography—there's not a lot of hard-core nuclear work there. In my defense, I would point out that I did pay attention when my National Laboratory colleagues "went nuclear." After 13 years at Los Alamos and Oak Ridge, some of it rubs off. I would also argue that a book that wishes to bring nuclear science to a broader audience must be written by someone who originally comes from outside the field and can still see it from an external perspective. My successful experiences teaching nuclear science to those that have a limited math and science background suggests that I might be a good choice to write this book. Ultimately, it will be the readers who decide if I have been successful in this effort.

This book draws heavily on those that have proceeded it, especially Ehmann and Vance's *Radiochemistry and Nuclear Methods of Analysis*, and Choppin, Rydberg, and Liljenzin's *Radiochemistry and Nuclear Chemistry*. It is not meant as a replacement to either, as they are both much more detailed; rather it is a contemporary resource for those wishing to learn a little bit of nuclear science. For those that wish to explore nuclear science beyond this text, a complete listing can be found in the bibliography.

I had often considered writing this textbook over the past five years of teaching nuclear chemistry and physics at the University of Wisconsin—La Crosse. It did not become a reality until Lance Wobus of Taylor & Francis asked to meet with me at a meeting of the American Chemical Society about a year and a half ago. I am very grateful to Lance for seeking me out, and for his constant support during the writing process.

This is easily the most ambitious writing project I have undertaken, and it would not have been possible without the assistance and encouragement of many others. First, I must acknowledge my predecessor in my current position at the University of Wisconsin—La Crosse, Willie Nieckarz. Based on his background in the nuclear navy, he pioneered our nuclear science programs. My nuclear chemistry course remains strongly based on his materials. I also acknowledge my students, who have never been reluctant to question me beyond the limits of my knowledge, and for their critical reading of earlier drafts of this text. I would especially like to single out former students Stephanie Rice, Rachel Borgen, and Jim Ironside for their extensive suggestions, and help preparing the solutions manual and figures. I must also thank my colleagues in the Departments of Chemistry and Physics at the University of Wisconsin—La Crosse; without their support and encouragement, this book would not have been written. I would also like to acknowledge my former colleagues and teachers who took the time to introduce me to nuclear science. Without them,

I would not have been hired into the wonderful job I now enjoy. Finally, I need to thank my family. Not only have they endured my many absences while preparing this text, but my father (William Bryan Sr., formerly an engineer who worked on the nuclear rocket program NERVA) and son (Lars Bryan, currently a sophomore studying mathematics at the University of Wisconsin—Madison) have read much of the text and provided numerous helpful suggestions. It is rare that three family generations have the opportunity to work together on a project like this.

Jeff C. Bryan
La Crosse, Wisconsin

Author

Jeff C. Bryan was born in Minneapolis in 1959, the second of four children. His arrival coincided with the purchase of a clothes dryer, and was therefore somewhat joyous—no more freeze-dried diapers during the long Minnesota winters. After two winters of being bundled up and thrown out into the snow to play with his brother, the family moved to Sacramento. Dr. Bryan was raised in Northern California during the 1960s and 1970s on hot-dish, Jell-o™[1] salad, reticence, and hard work. The conflict of the free and open society with the more reserved home life is the cause of many of Dr. Bryan's current personality disorders.

Dr. Bryan's father felt it was important that his children pay for their own way for their postsecondary education, thereby gaining an appreciation for advanced learning as well as being able to budget. Dr. Bryan decided to stay at home for his first two years of college, attending American River College in Sacramento. This was fortuitous, as he was able to learn chemistry from excellent teachers in a small classroom and laboratory-intensive environment. Dr. Bryan later earned a Bachelor of Arts degree in chemistry from the University of California, Berkeley, and his doctorate in Inorganic Chemistry from the University of Washington, Seattle, under the supervision of Professor Jim Mayer. His thesis presented a new chemical reaction, the oxidative addition of multiple bonds to low-valent tungsten.

After one year of postdoctoral work with Professor Warren Roper at Auckland University, Dr. Bryan spent five years at Los Alamos National Laboratory, initially as a postdoctoral fellow, then as a staff member. At Los Alamos his mentor, Dr. Al Sattelberger, encouraged him to begin research into the chemistry of technetium, which turned out to be modestly successful. Dr. Bryan then accepted a staff position at Oak Ridge National Laboratory, working with Dr. Bruce Moyer's Chemical Separations Group as the group's crystallographer. During this time there was a great deal of work done in the group investigating the separation of ^{137}Cs from defense wastes.

In 2002, Dr. Bryan made a long-anticipated move to academia by accepting a faculty position at the University of Wisconsin—La Crosse. He was hired to teach nuclear and general chemistry courses, primarily on the basis of his 13 years of research experience in the National Laboratories. Apparently the chemistry faculty was in desperate need to fill this position, as this background did not adequately prepare him for this dramatic change. In spite of the odds, this arrangement has worked out well for all concerned. It has gone so well that Dr. Bryan was promoted to Associate Professor in 2006 and was granted tenure in 2008. He also

---

[1] Jell-o is a delicious dessert and a registered trademark of the Kraft Foods Inc., Northfield, Illinois.

began teaching two physics courses, Radiation Physics and Introduction to Nuclear Science, in 2005.

Dr. Bryan has over 100 publications in peer-reviewed journals, including one book chapter and an encyclopedia article. His research has been featured on the covers of one book, two journals, and two professional society newsletters.

Dr. Bryan married the love of his life, Chris, in 1983. They are still happily married and are very proud of their three children, Katrina, Lars, and Tom.

1 Introduction

1.1 RADIATION

Radiation is scary. We can't see, hear, smell, or touch it, and we know that it can do terrible things to us. In large enough doses, it can kill us. Lower doses, especially if received over a long time, can increase our odds of getting cancer. It is easy to fear. Radiation is also well studied, and this knowledge can be used to dispel some of the fear it so easily engenders. This chapter will focus on the fundamentals of physics that can help us answer questions like "what is radiation," "where does it come from," and "what does it want with us?"

Much of this chapter will be devoted to definitions. In a sense this chapter is a foreign language tutorial where you'll learn to speak some nuclear science. Let's start with *radiation*. Radiation is energy that moves. More scientifically, it is energy propagated through a medium. There are all kinds of radiation all around us, all of the time. Radio stations use radio waves to transmit audio signals to our homes and cars. We also use radio waves to transmit signals to and from our growing number of wireless devices, such as cell phones and networked computers. Remote controls also use infrared light to control TVs, CD players, and other video or audio equipment. We use microwaves to transmit energy to our food. Visible light is also a form of radiation. Green light generated by the sun travels through space and our atmosphere, bounces off a plant, and travels through some more air to our eyes, which tell us the plant is green.

All of the examples above use photons to transmit their energy. Photons are commonly thought of as individual packets of visible light, but are also handy at transmitting all kinds of energy, as shown in figure 1.1. The type of energy they propagate depends on the wavelength (λ) and frequency (ν) of the photon. Note that the wavelength decreases while frequency increases from left to right. The product of wavelength and frequency is a constant, the speed of light ($c = 2.998 \times 10^8$ m/s):

$$c = \lambda \nu \tag{1.1}$$

The amount of energy each photon carries is directly related to the frequency:

$$E = h\nu \tag{1.2}$$

where h is Planck's constant (6.626×10^{-34} J.s). Therefore, electromagnetic radiation with high frequency and short wavelength, such as X-rays and γ-rays, is also carried by the highest-energy photons known, while radio frequencies are relatively low energy. Being more energetic has its consequences, as we shall see later in this chapter.

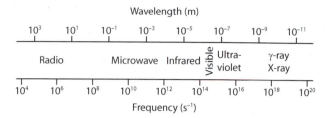

FIGURE 1.1   The electromagnetic spectrum.

1.2   ATOMIC STRUCTURE

Before we get into the more energetic forms of radiation, we need to know a bit about how atoms are put together, since most of the radiation we'll be concerned with in this book has its origin in the nuclei of atoms (**radioactivity**). Atoms are made up of a small, positively charged nucleus surrounded by a relatively large space (orbitals) occupied by tiny, fast-moving, negatively charged **electrons**. The nucleus is about 10,000 times smaller than the atom, which is defined by the approximate boundaries of the electron orbitals. This means that if the nucleus were the size of a marble, the atom would be as big as a football stadium! Atoms and all the objects that they make up, like this book, bananas, and people, are therefore mostly empty space. Despite its relatively small size, the nucleus accounts for over 99.9% of the mass of the atom, which means that almost all of the mass of an atom occupies very little space within the atom.

In this book, we'll use the Bohr model to describe the electronic structure of the atom. As shown in figure 1.2, the Bohr model shows the electrons orbiting the nucleus in circles. Readers should know that this is not terribly accurate. In reality, electrons occupy three-dimensional spaces called orbitals. These spaces are not exactly delineated since there is always a finite (however small!) probability of finding the electron just a little further from the nucleus. The Bohr model works just fine for us, because the circular electron orbits nicely represent the average distance between the electron and nucleus. Note also that the nucleus is not drawn to scale—if it were, it would be so small that we wouldn't be able to see it on the page.

The nucleus is made up of positively charged **protons** and neutral **neutrons**. Both have a mass of approximately 1 atomic mass unit or u ($1\,\text{u} = 1.66 \times 10^{-24}$ g), while the electron's mass is 5.5×10^{-4} u. The number of protons in a nucleus is equal to the atom's atomic number and determines what element it is. For example, an atom with eight protons in its nucleus is an oxygen atom, the eighth element listed in the periodic table. It is an oxygen atom regardless of the number of neutrons present. The number of neutrons will simply determine the mass of the atom and which **isotope** of oxygen this atom is.

If both the number of protons and the number of neutrons vary, we'll need a more general term than isotope. The term **nuclide** refers to any isotope of any element—a nucleus with any number of protons and neutrons. We can now define isotope a bit more efficiently as nuclides with the same number of protons, but different numbers of neutrons. Any nuclide can be represented using a nuclide symbol:

$$_Z^A X_N$$

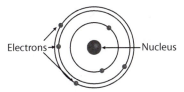

FIGURE 1.2   The Bohr atomic model.

where X is the elemental symbol from the periodic table, A is the **mass number** (number of protons plus neutrons), Z is the **atomic number** (number of protons), and N is the **neutron number** (number of neutrons). Since the mass number is equal to the number of protons plus the number of neutrons.

$$A = Z + N \tag{1.3}$$

It is fairly rare that the neutron number is included in a nuclide symbol. To find the number of neutrons, one only needs to subtract the number of protons from the mass number. Also, the atomic number is sometimes left out of a nuclide symbol, since the elemental symbol can also give us that value. When spoken aloud, the nuclide symbol is given as "element-A." For example, ^{99}Tc would be said as "technetium-99." Technetium-99 is another way to represent a nuclide symbol in writing. Finally, ^{99}Tc can also be represented as "Tc-99." These other methods of representing a nuclide are likely holdovers from the days when it was difficult to place superscripts into manuscripts.

Just as isotopes have a constant number of protons, isotones must all have the same number of neutrons. More formally, **isotones** are nuclides with the same number of neutrons, but different numbers of protons. **Isobars** are nuclides with the same mass number, but differing numbers of protons and neutrons. As examples, consider the following nuclides:

$$^{16}_{8}O_{8} \qquad ^{18}_{8}O_{10} \qquad ^{15}_{7}N_{8} \qquad ^{18}_{9}F_{9}$$

Of these, only ^{16}O and ^{18}O are isotopes, ^{16}O and ^{15}N are isotones, and ^{18}O and ^{18}F are isobars.

Collectively, neutrons and protons are known as **nucleons**. Can nucleons be further subdivided into even smaller particles? Yes, both are partially made up of **quarks**. There are six types, or flavors, of quarks, such as up, down, top, bottom, charmed, and strange. We will only be concerned with up quarks, which have a $+2/3$ charge, and down quarks, which have a $-1/3$ charge. Protons have two up quarks, and one down quark, for a net charge of $+1$. Neutrons are made up of one up quark and two down quarks, for a net charge of 0. It is interesting to note that the difference between the two is a single quark. We can also imagine that interconversion between neutrons and protons can be thought of as the flipping of a quark, from down to up or up to down.

So what holds the nucleus together? At first glance it looks like an impossibility—all those positive charges in a relatively small space should push themselves apart. The force pushing them apart is electrostatic: like charges repel each other and opposite charges attract. This is also known as Coulomb repulsion (or attraction), which acts over fairly long distances. An attractive force known as the nuclear **strong force** acts between the nucleons within the nucleus, but it has a very short range. It can only act over ~10^{-15} m. The average nucleus is ~10^{-14} m in diameter, so this is a force that has a very limited range, even in a really small space like a nucleus. Even though it acts over a very short distance, the strong force is more powerful (~100×!) than the Coulomb repulsion threatening to rip the nucleus apart. It is the strong force that holds the nucleons together in the nucleus.

1.3 NUCLEAR TRANSFORMATIONS

Now that we have an idea of what nuclei are made of, why they hold together, and how to represent them as symbols, let's start messing with them. Fundamentally, there are two types of nuclear transformations: decay and reaction. This is wonderful simplicity when compared to the world of chemistry where the numbers and types of chemical reactions are constantly proliferating. However, it is an oddity of nuclear science that decay processes are not referred to as "reactions." Nuclear reactions, which we'll study in more detail in Chapter 7, are only those transformations where a particle (like a neutron or a ^{4}He nucleus) or a photon interacts with a nucleus, resulting in the formation of another nuclide and some other stuff. However, we're getting ahead of ourselves.

The first main transformation type we'll consider is nuclear decay. **Decay** starts with an unstable nuclide that typically spits out a particle and/or a photon while turning itself into something more stable. An example would be ^{222}Rn decaying to ^{218}Po by ejecting two neutrons and two protons all at once: a $^4_2$He nucleus, also known as an **alpha particle**:

$$^{222}_{86}\text{Rn} \rightarrow \, ^{218}_{84}\text{Po} + ^4_2\text{He}$$

This is an example of **alpha decay**, or it could be said that ^{222}Rn undergoes alpha decay. Notice that the atomic and mass numbers are the same on both sides of the arrow. ^{222}Rn starts out with 86 protons and 222 total nucleons. Together, ^{218}Po and an alpha particle have 86 protons and 222 total nucleons (check this for yourself!). This is how nuclear equations are "balanced," by simply comparing the mass numbers (superscripts) and atomic numbers (subscripts). They should always add up in the same way on either side of the arrow.

There are two other basic types of nuclear decay: **beta** and **gamma**. Additionally, there are three different kinds of beta decay. The first is often referred to simply as "**beta decay**," but is sometimes called "beta minus" or "negatron" decay. This form of beta decay involves the ejection of an electron from the nucleus—a statement that shouldn't make much sense! Electrons are supposed to spend their time in the space around the nucleus, so how can one be emitted *by* the nucleus? Let's take a look at an example and see if it can be explained. ^{14}C decays to ^{14}N by spitting out an electron, also known as a **beta particle**:

$$^{14}_{6}\text{C} \rightarrow \, ^{14}_{7}\text{N} + \, ^0_{-1}e$$

The beta particle is symbolized here as $_{-1}^{0}e$, but is sometimes represented as $_{-1}^{0}\beta$ or as β^-. The mass number for an electron should be equal to zero because it contains no protons or neutrons. The "atomic number" of -1 simply accounts for its negative charge. It also balances our nuclear decay equation quite nicely.

How does this electron come from the nucleus? Look at what happens to the numbers of protons and neutrons in this decay. The carbon nuclide has six protons and eight neutrons, while the nitrogen nuclide has seven protons and seven neutrons. A neutron has turned itself into a proton in this decay (flipped a quark):

$$_{0}^{1}n \rightarrow\ _{1}^{1}p +\ _{-1}^{0}e$$

When that positive charge is created, a negative charge must also be created, for balance. We can therefore think of a neutron converting to a proton and an electron within the ^{14}C nucleus. The proton in the decay above can also be symbolized as $_{1}^{1}$H, a hydrogen-1 nucleus.

The second kind of beta decay is known as **positron** or "beta plus" decay. In this type of decay a proton is converted into a neutron and a positron. A positron is a particle with the same mass as an electron, but with a positive charge. An example is the decay of ^{18}F to ^{18}O:

$$_{9}^{18}\text{F} \rightarrow\ _{8}^{18}\text{O} +\ _{+1}^{0}e$$

Positrons can also be symbolized as $_{+1}^{0}\beta$ or β^+. Positrons are a form of antimatter (the same stuff that fuels the starships in Star Trek!). Every particle that we know about in our universe, such as electrons, has a corresponding antiparticle. In this case, the difference between particle and antiparticle is simply charge.

The final type of beta decay is **electron capture**. Just like positron decay, electron capture involves the conversion of a proton into a neutron, but it's done a little differently. As its name implies, one of the electrons surrounding the nucleus is drawn into the nucleus, transforming a proton to a neutron. An example is the decay of ^{67}Ga:

$$_{31}^{67}\text{Ga} +\ _{-1}^{0}e \rightarrow\ _{30}^{67}\text{Zn}$$

The words "orbital electron" are often added below the electron in the decay equation above to indicate this particle does not originate in the nucleus. In other words, we should not think of it as a beta particle, nor should it ever be represented with the Greek letter β. Electron capture is unique among all forms of nuclear decay in that it has more than the unstable nuclide on the left-hand side of the arrow. We can excuse this poor behavior because the electron is not a nuclear particle.

Notice that this is the first time since atomic structure was described earlier in this chapter that the atom's (orbital) electrons have been mentioned. Generally speaking, this textbook is not concerned with the electrons or even the chemical form of the atom. It is instead focused on the nucleus and nuclear changes.

Our final type of nuclear decay is **gamma decay**. Gamma decay is unique in that no particles are emitted or absorbed, and both the numbers of protons and neutrons remain constant. It is simply the rearrangement of the nucleons in a nucleus. Initially, the nucleons are arranged poorly energetically—one or more nucleons occupy a higher energy state than they need to, and the nucleus is said to be in an excited state. Rearrangement allows the nucleons that are in unusually high energy states to drop down into lower ones. The lowest energy configuration would be called the ground state. Rearrangement from a higher energy state to a lower one requires the release of energy, which is done in the form of a high-energy photon, or γ-ray. An example is the decay of ^{99}Tc* to ^{99}Tc:

$$^{99}_{43}\text{Tc}* \rightarrow \, ^{99}_{43}\text{Tc} + \gamma$$

The asterisk indicates that the nuclide to the left of the arrow is in an excited state. Two nuclides that differ only in energy state (such as ^{99}Tc* and ^{99}Tc) are called **isomers**.[1] The gamma photon can also be represented as $^{0}_{0}\gamma$, but the zeroes are usually left off. Being a photon, it has no protons, neutrons, or electrical charge.

Some excited (isomeric) states exist for reasonable amounts of time. If it hangs around for more than a second or two, it is often referred to as a **metastable** (or simply "meta") state, and a lowercase m is added after the mass number in the nuclide symbol, instead of the asterisk following the element symbol. ^{99}Tc has a metastate, which is symbolized as ^{99m}Tc.

Nuclear reaction is the other main kind of nuclear transformation we'll study. It involves the collision of a particle and a nucleus (or two nuclei) with the formation of a product nuclide and either a particle or photon. An example would be a neutron colliding with a ^{98}Mo nucleus to form ^{99}Mo and a gamma photon:

$$^{98}_{42}\text{Mo} + \, ^{1}_{0}\text{n} \rightarrow \, ^{99}_{42}\text{Mo} + \gamma$$

In this reaction, the neutron is the **projectile**, the ^{98}Mo is the **target**, and the ^{99}Mo is the **product**. The easy way to distinguish between decay and nuclear reactions is that reactions always have at least two nuclides, particles or photons on either side of the arrow. Decay equations generally (except electron capture) only have a nuclide on the left side. Like all of the nuclear equations in this section, this one is balanced, both in terms of atomic and mass numbers. There are special kinds of nuclear reactions, but we'll worry about them later.

1.4 NUCLEAR STABILITY

A **radioactive** nuclide is one that spontaneously undergoes nuclear decay. Why do some nuclides decay, but others do not? The question is really one of nuclear

---

[1] As noted in Chapter 4, many nuclear science textbooks define isomers as only those excited nuclear states with measurable lifetimes. Because of the arbitrary nature of this definition, isomers are defined more broadly here.

stability, and the answer depends on numbers. Perhaps the most important number related to nuclear stability is the neutron-to-proton ratio (N/Z). If we plot the number of protons (Z) vs. the number of neutrons (N), and make a mark for every stable nuclide known, the result will look like figure 1.3. The region where stable nuclides are almost always found is called the **belt of stability**. Notice that for low Z nuclides (near the top of the periodic table), the belt is narrow and is very close to 1:1. It is difficult to see on this figure, but there are only two stable nuclides with an N/Z ratio <1: ^{1}H and ^{3}He. As Z increases, the belt widens, and the belt's N/Z ratio also increases.

How can we use this information? If a nuclide's N/Z ratio is too high or too low, it will not lie within the belt, and odds are very good it will not be stable. Lying outside of the belt means that the nuclide has too many protons or too many neutrons to be stable. Therefore, it will likely undergo some type of beta decay to an isobar. A large number of known nuclides lie off the belt, so examination of the N/Z ratio is an important method to determine nuclear instability.

The belt is not completely filled with dots, telling us that there are a number of unstable nuclides with good N/Z ratios. Why? Again, we need to look at some numbers. If we take all the known *stable* nuclides and count how many have an even number of protons (Z) and an even number of neutrons (N), there are about 159. Statistically speaking there should be equal numbers of stable nuclides with an even Z and odd N, odd Z and even N, or odd Z and odd N. Curiously enough, this is not true. As can be seen in table 1.1, if either Z or N is odd and the other is even, then there are only about 50 stable nuclides. If *both* Z and N are odd, there are only four! The four exceptions are $^2_1$H$_1$, $^6_3$Li$_3$, $^{10}_5$B$_5$, and $^{14}_7$N$_7$. Notice that all have a low number of protons. Even is good and odd is bad, but why? Apparently protons and neutrons like to pair up with others of their kind, and this provides

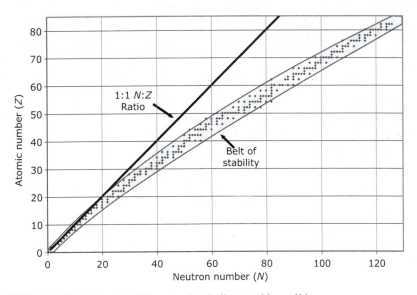

FIGURE 1.3 The belt of stability—the dots indicate stable nuclides.

TABLE 1.1

Stable Nuclides

Z	N	#Stable
Even	Even	159
Even	Odd	53
Odd	Even	50
Odd	Odd	4

stability. An odd/odd nuclide can become even/even by converting a proton to a neutron or a neutron to a proton. So, some type of beta decay looks likely for odd/odd nuclides.

Why does figure 1.3 stop at $Z = 80$? Beyond element 82 (Pb) there are no stable nuclides. They are simply too big for the very short-ranged strong force to hold the nucleus together against the Coulomb repulsion of all those protons. Nuclides this large *tend* to undergo alpha decay, as this lowers both Z and N.

Finally, close examination of figure 1.3 reveals that some dots seem to lie just outside the belt of stability. All the rationalizations (bad N/Z, odd/odd, too big) we've seen so far are different ways to understand *instability*. How can we rationalize unusual *stability*? We'll need more information first. Are there other instances where unusual stability is observed? Yes, certain elements seem to have an unusually large number of stable isotopes. Most elements only have 1, 2 or 3 stable isotopes, but lead has 4, nickel has 5, calcium has 6, and tin has 10! There is something special about certain numbers of protons or neutrons. Detailed analysis reveals that the numbers 2, 8, 20, 28, 50, 82, and 126 seem to impart special stability to nuclides that contain these numbers of protons and/or neutrons. Because of this characteristic, they are sometimes referred to as **magic numbers**.

Far from having mystical qualities, these numbers have a great analogy in chemistry. Students of chemistry know that chemical stability is imparted on those elements occupying the far right column of the periodic table, the so-called noble gases. This stability comes from exactly filling their valence shells with electrons. In chemistry, these numbers would be 2, 8, 18, and 32. They are not referred to as "magic," but they could be. Table 1.2 lists all stable nuclides containing magic numbers of nucleons. Note that a few have magic numbers of both protons *and* neutrons. These nuclides can be expected to be extra stable compared to others with similar numbers of nucleons.

1.5   IONIZING RADIATION

This chapter started by discussing radiation as energy propagated through a medium, but stayed focused on electromagnetic radiation (photons). We have since learned about radioactive decay, and know that this happens in order to achieve greater nuclear stability—therefore energy must be released as part of all nuclear decay processes.

TABLE 1.2

Stable Nuclides with Magic Numbers of Nucleons

2	8	20	28	50	82	126
			Neutrons			
^{4}He	^{15}N	^{36}S	^{48}Ca	^{86}Kr	^{136}Xe	^{208}Pb
	^{16}O	^{37}Cl	^{50}Ti	^{88}Sr	^{138}Ba	
		^{38}Ar	^{51}V	^{89}Y	^{139}La	
		^{39}K	^{52}Cr	^{90}Zr	^{140}Ce	
		^{40}Ca	^{54}Fe	^{92}Mo	^{141}Pr	
					^{142}Nd	
					^{144}Sm	
			Protons			
^{3}He	^{16}O	^{40}Ca	^{58}Ni	^{112}Sn	^{204}Pb	
^{4}He	^{17}O	^{42}Ca	^{60}Ni	^{114}Sn	^{206}Pb	
	^{18}O	^{43}Ca	^{61}Ni	^{115}Sn	^{207}Pb	
		^{44}Ca	^{62}Ni	^{116}Sn	^{208}Pb	
		^{46}Ca	^{64}Ni	^{117}Sn		
		^{48}Ca		^{118}Sn		
				^{119}Sn		
				^{120}Sn		
				^{122}Sn		
				^{124}Sn		

Ionizing radiation is when a particle or a photon has enough energy to turn an unsuspecting atom into an ion. In other words, the radiation needs to have enough energy to knock an electron or a proton loose from an atom. The former process is much more probable, so much so that we'll ignore the latter.

On the electromagnetic spectrum, only X-rays and γ-rays have sufficient energy to ionize atoms. All alpha and beta particles produced in nuclear decay also have enough energy to ionize matter. Therefore X-rays, γ-rays, alpha particles, and beta particles can all be considered forms of ionizing radiation.

Unfortunately, the more general term "radiation" is commonly, but incorrectly, used to refer to the energetic photons and particles emitted as part of nuclear decay processes. This text will strive to use the terms "radiation" and "ionizing radiation" properly. The reader should know that the misusage of "radiation" is very common, and even those that know better will sometimes slip.

Even though the three major forms of decay, alpha, beta, and gamma, can all create ions when their emitted particles or photons interact with matter, their relative abilities to penetrate matter differ greatly. Alpha particles have a high probability of interacting with matter. They carry a 2+ charge and, relative to electrons and photons, they are very large. Alpha particles can be effectively stopped by thin pieces of paper, or by just a few centimeters of air. Beta particles are electrons, so they have a lot less mass and only half the charge of alpha, but still have a reasonably high probability of interacting with matter. They can generally be stopped by a centimeter or two of water. Energetic photons such as X-rays and γ-rays have no charge or mass, and therefore have a relatively low probability of interacting with matter. The best shielding material for them is high-density matter, such as lead.

1.6 A BIOLOGICAL THREAT?

One of the reasons humans fear ionizing radiation is that we know that, in large enough doses, it can hurt us. From what we've learned so far it shouldn't be hard to figure out how. Ionizations inside our bodies and possibly in our DNA can cause cells to die, or to replicate poorly. With enough of a dose, we start to observe measurable effects such as lower blood cell counts. As the dose increases, we know that the health effects gradually get worse until death becomes pretty much certain.

What about below the level where measurable health effects are known? Since the health effects cannot be measured, there is some uncertainty as to what low doses mean. Most believe that the linear relationship between risk and dose at higher exposure levels can be extrapolated through low doses, as illustrated in figure 1.4. This means that any additional exposure *could* lead to deleterious health effects. With this in mind, it is always a good idea to minimize dose due to exposure to ionizing radiation, when possible. This is the basis of the *As Low As Reasonably Achievable* (**ALARA**) policy followed by those who work with radioactive materials. If there's more than one way to do something, pick the path that gives a lower dose.

There is also ample evidence that our cells can repair or replace many of the cells that are damaged or killed due to exposure to ionizing radiation. We are all exposed to naturally occurring ionizing radiation every minute of every day. Some are exposed to more than others, such as folks that live at higher altitudes. There are no measurable health effects due to the extra exposure of the large populations of humans that live in these areas. This might suggest to some that our bodies can safely cope with low levels of ionizing radiation. A more prudent conclusion is that we have difficulty measuring health effects at these levels.

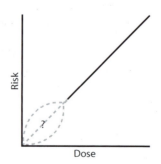

FIGURE 1.4   Dose vs. risk.

Naturally occurring ionizing radiation comes from several sources. First, some of the lighter elements present in the upper atmosphere (like N, O, and H) can undergo nuclear reactions that are a result of cosmic radiation. ^{3}H and ^{14}C are constantly produced in this way. These nuclides find their way back to earth and are incorporated into molecules that make up living things since their chemistries remain the same as other hydrogen and carbon atoms.

Another source comes from radioactive nuclides that take a very long time to decay. If that time is longer than the age of the earth (4.5×10^9 years), then they're still around us. All together, there are about 20 of them. One of the most notable is ^{40}K. Since all isotopes of an element have the same chemical behavior, and because we humans need potassium in our bodies, we all have some ^{40}K in our bodies. Anything that contains potassium (bananas, Gatorade®, your friends) is naturally radioactive.

All but 3 of these 20 long-lived radioactive nuclides decay to stable nuclides. The three that don't are ^{232}Th, ^{235}U, and ^{238}U. They all eventually decay to stable isotopes of lead, but along the way they produce a number of other naturally occurring radioactive nuclides, including ^{222}Rn. This is our final source of naturally occurring radioactive nuclides. The paths they take to Pb are called the naturally occurring **decay series** (or **chains**). To get there, a number of alpha and beta (β^-) decays need to take place, as illustrated in figure 1.5 for the series originating with ^{238}U. Notice that an alpha decay lowers both the proton and neutron numbers by 2, while a beta decay lowers N by 1, but increases Z by 1. At certain nuclides (like ^{218}Po), the path splits into two. This means that more than one decay pathway exists for that particular nuclide.

What about all the radioactive materials released by humans (anthropogenic sources)? Despite our best efforts, we've allowed a fair bit of radioactive

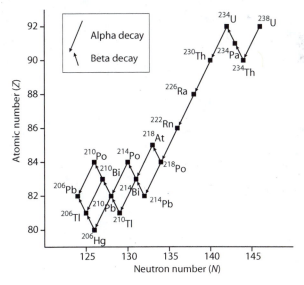

FIGURE 1.5   The ^{238}U decay series.

material to get into the environment. The following are some of the more significant sources:

The development, manufacture, testing, and use of nuclear weapons
Accidental releases from nuclear power plants such as Chernobyl (1986) and
  Three Mile Island (1979)
Regulated releases from nuclear power plants
Irresponsible disposal of radioactive waste
The manufacture, use, and disposal of radioactive nuclides for research or
  medical applications.

Sounds bad, but all five sources combined account for less than 1% of the average annual human dose in the United States. They are all tabulated together under "Other" in figure 1.6, which illustrates the major sources for average annual doses from ionizing radiation in the United States.

  The first four bars shown in figure 1.6 are from natural sources. By far, we get the most from radon. That's because it is a gas. Once formed in the decay series, it can get into our water or into the air we breathe. It gives us a large dose because we ingest it, and it can undergo alpha decay while inside us. Once emitted, the alpha particle cannot escape our body and ends up doing a lot of local damage. Don't stop breathing the air or drinking the water—your friends and relatives might miss you! Besides, radon has always been present on the planet, and humans have evolved with some ability to repair the damage it causes. Even so, we should keep the ALARA principle in mind. Since radon is responsible for our largest dose, we should take steps to minimize our exposure. Since a good chunk of this dose comes from the air we breathe at home, it is a good idea to test your home for radon. As we'll soon learn, you will get the best results from a longer test—choose one that is in your home for a few months or longer.

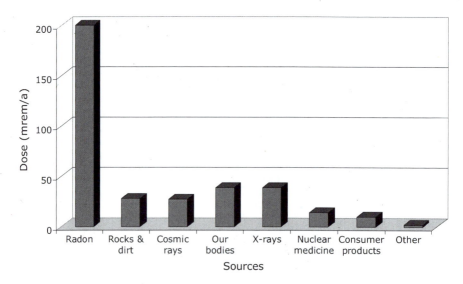

FIGURE 1.6   Average annual radiation doses in the United States.

N → Z ↓	0	1	2	3	4	5	6	7	8	9	10	11	12	13	14	15	16
8					O-12	O-13 17.77	O-14 5.143	O-15 2.754	O-16 99.757	O-17 0.038	O-18 0.205	O-19 4.821	O-20 3.814	O-21 8.11	O-22 6.5	O-23 11.3	O-24 11.4
7					N-11 2.290	N-12 17.338	N-13 2.2204	N-14 99.632	N-15 0.368	N-16 10.420	N-17 8.68	N-18 13.90	N-19 12.53	N-20 18.0	N-21 17.2	N-22 22.8	N-23 23
6			C-8 2.142	C-9 16.498	C-10 3.6478	C-11 1.983	C-12 98.93	C-13 1.07	C-14 0.15648	C-15 0.9772	C-16 8.011	C-17 13.17	C-18 11.81	C-19 17.0	C-20 15.8		C-22 21
5			B-7 2.200	B-8 17.979	B-9 0.277	B-10 19.9	B-11 80.1	B-12 13.369	B-13 13.437	B-14 20.64	B-15 19.09		B-17 22.7		B-19 26.5		
4			Be-6 1.372	Be-7 0.86182	Be-8 0.0918	Be-9 100	Be-10 0.556	Be-11 11.51	Be-12 11.71	Be-13 17.1	Be-14 16.2						
3		Li-4 3.100	Li-5 1.970	Li-6 7.59	Li-7 92.41	Li-8 16.0045	Li-9 13.606	Li-10 0.42	Li-11 20.62								
2		He-3 0.00014	He-4 99.9999	He-5 0.890	He-6 3.508	He-7 0.440	He-8 10.65	He-9 1.15	He-10 15.8								
1	H-1 99.9885	H-2 0.0115	H-3 0.18591	H-4 2.910	H-5 25	H-6 24.2											

FIGURE 1.7 A chart of the nuclides for the first eight elements.

Introduction to Nuclear Science

You may have expected cosmic rays to account for a larger percentage of our annual dose. Much of the cosmic radiation we're exposed to is in the form of high-energy photons, which have a very low probability of interacting with the matter we are made of. That's the difference between exposure and dose: our greatest exposure to ionizing radiation is cosmic, but our greatest dose comes from radon, because radon is inside us and emits ionizing radiation with a very high probability of interacting with matter, while cosmic rays have a relatively low probability of interacting.

The final four sources shown in figure 1.6 are anthropogenic. They are almost entirely due to medical diagnostic procedures. Remember, these bars are for the average U.S. citizen each year. Not everyone has an X-ray or nuclear medicine procedure every year—these numbers come from the dose for all procedures done each year divided by the total population.

1.7 THE CHART OF THE NUCLIDES

The chart is a valuable tool for anyone studying nuclear science. We've already seen a couple of extremely condensed versions of it in figures 1.2 and 1.4. A more conventional, albeit limited, representation is shown in figure 1.7. Basically, it is a way to provide information on all known nuclides, stable (white) and unstable (gray). It is laid out with N increasing from left to right (the x-axis) and Z increasing from bottom to top (y-axis). Thereby, each square represents a unique N and Z combination—a unique nuclide. Percent natural abundance is given at the bottom of the square for each stable nuclide and total energy (MeV, Chapter 3) given off during decay is given for all the unstable nuclides.

Printed versions are published by Lockheed-Martin Corporation and are available for purchase through http://www.chartofthenuclides.com/. The printed charts also come with extensive text explaining how to properly interpret the information provided for each nuclide. There are also excellent web-based versions of the chart. A fairly intuitive version is available at http://atom.kaeri.re.kr/ton/. Many of the questions following each chapter in this textbook will assume that the reader has access to the kinds of information available in the chart.

QUESTIONS

1. Which of the following statements are true?
   (a) All radiation is harmful.
   (b) All ionizing radiation is anthropogenic.
   (c) Atoms cannot be changed from one element to another.
   (d) You are exposed to ionizing radiation every minute.
   (e) The human body is naturally radioactive.
   (f) The human body is capable of detecting radioactivity at any level.
2. List some of the apprehensions you have or have had concerning ionizing radiation.
3. What is radiation? What is ionizing radiation and where does it come from?

4. What wavelength of photon has an energy of 7.84×10^{-18} J? Is this ionizing radiation? Briefly explain your answer.

5. What do γ-rays and UV light have in common? What distinguishes them?

6. Is the phrase 'nature abhors a vacuum' accurate based on our knowledge of atomic structure?

7. If you built a scale model of an atom, and used a baseball to represent the nucleus, how big would the atom have to be?

8. Define the following: nuclide, nucleon, beta particle, positron decay, the electromagnetic spectrum, isotope, isobar, and ALARA.

9. How many electrons could occupy the same space as an alpha particle?

10. Complete and balance the following. Identify each as decay or nuclear reaction. If it is decay, give the type of decay.

$$^{81}_{37}\text{Rb} \rightarrow {}^{81}_{36}\text{Kr} + \qquad {}^{14}_{7}\text{N} + \rightarrow {}^{1}_{1}\text{H} + {}^{14}_{6}\text{C} \qquad \rightarrow {}^{137}_{56}\text{Ba} + {}^{0}_{-1}e$$

$$^{239}_{94}\text{Pu} \rightarrow {}^{4}_{2}\text{He} + \qquad {}^{201}_{80}\text{Hg} + {}^{2}_{1}\text{H} \rightarrow 2{}^{1}_{0}n + \qquad {}^{208}_{83}\text{Bi} + {}^{0}_{-1}e \rightarrow$$

11. Why are ^{32}P, ^{44}Ti, ^{20}O, ^{100}Ru*, and ^{212}Po unstable nuclides? Write out their decay equations.

12. ^{48}Ca shouldn't be a stable nuclide, but is. Explain this contradiction.

13. One of the naturally occurring decay series begins with ^{232}Th and ends with ^{208}Pb. What is the minimum number of alpha and beta decays required for this series?

14. A fourth naturally occurring radioactive decay series starting with ^{237}Np may have existed in the past. Using a chart of the nuclides, graph this decay series (similar to fig. 1.4) and explain why it does not exist today.

15. Using figure 1.6, estimate the percentage of the average annual U.S. dose due to natural sources.

16. List at least two anthropogenic sources of radioactive nuclides now found in the environment. Do a little research outside of this textbook and explain how these nuclides were (are) released.

17. Why are so many more neutron-rich nuclides shown in figure 1.7 than proton-rich nuclides?

2 The Mathematics of Radioactive Decay

2.1 ATOMIC MASSES AND AVERAGE ATOMIC MASSES

Before looking at the mathematics of decay, it is important to remember the difference between atomic mass and average atomic mass. **Atomic mass** is the mass of a single atom. The periodic table lists the average atomic mass for each element. Its units can be either atomic mass units (u, a.k.a the unified atomic mass unit, sometimes abbreviated as amu) or grams per mole. Atomic mass units refer to the mass of a single atom and are more commonly used in nuclear science. One u is equal to 1.66054×10^{-24} g. **Average atomic mass** is a weighted average of the masses of each naturally occurring isotope for a particular element. "Weighted" means that if one isotope has a greater natural abundance for a particular element, it contributes more significantly to the average atomic mass.

Example: Calculate the average atomic mass of copper.

The percent natural abundances and masses of the two naturally occurring isotopes of Cu are available in the chart of the nuclides. They are organized in table 2.1.

To determine the average atomic mass, add together the products of the atomic masses and the *fractional* abundances.

$$(62.9296 \text{ u} \times 0.6917) + (64.9278 \text{ u} \times 0.3083) = 63.54 \text{ u}$$

Notice that the average atomic mass is closer to 63 u than it is to 65 u.

2.2 THE NATURE OF DECAY

Since humans can't sense radioactive decay, we need some other way to measure it. There are a variety of radiation detectors, as discussed in Chapter 6. For now, we just need to know that it can be done, albeit not always with high efficiency. Radioactive decay is **isotropic**, which means that particles and/or photons are emitted equally in all directions by a source of radioactivity. Detectors only detect radiation that interacts with them, typically a small fraction of all radiation being emitted at the time. Data recorded by detectors are given units of count rate, such as counts per minute (cpm). If all of the decays from a source are counted over time, then the decay rate, or **activity**, is being measured. Decay rate is often given in units of decays per minute (dpm) or decays per second (dps). Those who work with radioactive materials often use count rate and decay rate interchangeably. This is understandable, since they are proportional to each other. Readers of this book would do well to know the difference.

TABLE 2.1

Mass and Abundance Data for Copper

Isotope	% Abundance	Atomic mass (u)
^{63}Cu	69.17	62.9296
^{65}Cu	30.83	64.9278

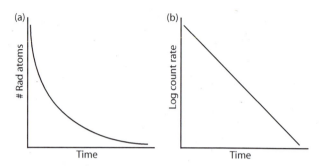

FIGURE 2.1 (a) The change in count rate over time for radioactive decay. (b) The change in the logarithm of the count rate with time.

Over time, as a radioactive nuclide decays, it becomes less radioactive. We could make measurements of this change with a detector and a clock. Once we've collected these data, it could be plotted as shown in figure 2.1a. Mathematically, radioactive decay is an exponential process, meaning that for equal intervals of time, the count rate will decrease by an equal percentage. This also means that if instead of plotting count rate vs. time, we plot the logarithm of the count rate vs. time, we'll get a straight line, as shown in figure 2.1b.

Two different forms of the same mathematical equation can describe the change in radioactivity of a source:

$$\ln \frac{A_2}{A_1} = -kt \tag{2.1}$$

$$A_2 = A_1 e^{-kt} \tag{2.2}$$

where A represents activity, or count rate so long as it is applied consistently in one equation. A_1 is the activity at the beginning of time t and A_2 is the activity at the end of that time. ln is the natural logarithm function and e is natural antilogarithm. k is the **decay constant** and has units of reciprocal time (time^{-1}). The decay constant is an indication of how quickly a source will decay, and is sometimes represented by λ. Because of the awkward units associated with the decay constant, it is often converted to half-life ($t_{1/2}$):

$$k = \frac{\ln 2}{t_{1/2}} \approx \frac{0.693}{t_{1/2}} \tag{2.3}$$

Half-life is defined as the amount of time it takes for the decay rate (or count rate) to reach half of its original value. If a nuclide starts out with an activity of 1000 dpm, it will be 500 dpm after one half-life, 250 after two, and so on. Half-life is commonly used in nuclear science. A nuclide with a very long half-life will decay slowly and be radioactive for a very long time. In contrast, a nuclide with a short half-life will decay quickly.

Example: A radionuclide has a half-life of 6.01 h. If its activity is currently 5430 dpm, what will its activity be 4.00 h from now?

$$A_2 = A_1 e^{-kt} = A_1 e^{-[(\ln 2)/t_{1/2}]t} = 5430 \text{ dpm} \times e^{-[(\ln 2)/6.01 \text{ h}] \times 4.00 \text{ h}} = 3423 \text{ dpm}$$

Less than one half-life has gone by, so we expect the answer to be greater than half the original. Notice also that the time units need to match within the exponent (both are hours), and the activity units for A_1 and A_2 must also match. If your instructor is a stickler for significant figures, then the above answer should be rounded to 3420 dpm. Generally speaking, three significant figures are the best that can be expected from detector measurements.

Example: A radioactive nuclide had an activity of 1.38×10^5 dpm exactly 60 days ago, but now has an activity of 6.05×10^4 dpm. What is its half-life?

$$\ln \frac{A_2}{A_1} = -kt$$

$$\ln \left(\frac{6.05 \times 10^4 \text{ dpm}}{1.38 \times 10^5 \text{ dpm}} \right) = -\frac{0.693}{t_{1/2}} (60 \text{ d})$$

$$t_{1/2} = 50.4 \text{ d}$$

In this example, approximately one half-life has passed, and the activities differ by (roughly) a factor of two. Notice that units for $t_{1/2}$ come from t.

Activity is also commonly expressed in units of curies (Ci) or becquerels (Bq). The curie is named after Marie Curie, the Polish physicist who performed much of the early groundbreaking work with radioactive materials. One curie is defined as exactly 3.7×10^{10} dps or the approximate rate of decay of 1 g of ^{226}Ra. This is a rather large number, so millicuries (mCi) or microcuries (μCi, 1 μCi $= 2.22 \times 10^6$ dpm) are often used for smaller sources. Remember that the milli prefix means 10^{-3} and micro means 10^{-6}. The becquerel (Bq) is named after Henri Becquerel, the French physicist who is credited with the discovery of radioactivity. It is the *Système International d'Unités* (SI) unit and is equal to 1 dps. This is a tiny value, so megabecquerels (MBq, 10^6 Bq) are often used.

Example: A ^{90}Sr source was calibrated to emit 1.00 μCi of radiation. If its activity today is measured at 5.76×10^5 dpm, how long ago was it calibrated?

$$1.00 \text{ } \mu\text{Ci} \times \left(\frac{2.22 \times 10^6 \text{ dpm}}{1 \text{ } \mu\text{Ci}} \right) = 2.22 \times 10^6 \text{ dpm}$$

$$\ln \frac{5.76 \times 10^5 \text{ dpm}}{2.22 \times 10^6 \text{ dpm}} = -\frac{\ln 2}{28.78 \text{ a}} t$$

$$t = 56 \text{ a}$$

In this example, it's necessary to convert one of the two given activities to match the units of the other. The choice to convert A_1 to dpm was arbitrary. A_2 could've been converted to μCi instead. It's also necessary to look up the half-life for ^{90}Sr (28.78 years). The abbreviation for years is a, from the Latin word for year, *annum*. Finally, notice that roughly two half-lives have passed, and the activity is roughly 1/4 of its original value.

As mentioned at the beginning of the chapter, detectors typically only 'see' a fraction of all decays. The percentage of all decays observed is called the **percent efficiency**:

$$\% \text{ efficiency} = \frac{\text{cpm}}{\text{dpm}} \times 100\% \qquad (2.4)$$

Example: A ^{210}Pb source was calibrated at 0.0202 μCi 31.0 years ago. Under your detector it records 2.16×10^3 cpm. What is your detector's percent efficiency?

First calculate the activity in dpm:

$$0.0202 \text{ μ Ci} \times \frac{2.22 \times 10^6 \text{ dpm}}{\text{μ Ci}} = 4.48 \times 10^4 \text{ dpm}$$

After looking up the half-life of ^{210}Pb, calculate what the *activity* is today:

$$A_2 = 4.48 \times 10^4 e^{-(0.693/22.3 \text{ a})31.0 \text{ a}} = 1.71 \times 10^4 \text{ dpm}$$

Finally, use the count data to determine the efficiency:

$$\% \text{ efficiency} = \frac{\text{cpm}}{\text{dpm}} \times 100\% = \frac{2.16 \times 10^3 \text{ cpm}}{1.71 \times 10^4 \text{ dpm}} \times 100\% = 12.6\%$$

In addition to the limited size of the detector, there are a number of experimental conditions that can affect the percent efficiency and, therefore, the observed count rate. These problems are discussed in Chapter 6.

So long as the conditions of the experiment remain constant, the count rate will be proportional to activity, and either can be used in the equations presented at the beginning of this chapter. There's another important proportionality that needs to be considered. If there's more radioactive material, there should be more activity. The mathematical relationship is remarkably simple:

$$A = kN \qquad (2.5)$$

where A is activity, k is the decay constant, and N is the number of atoms of the radioactive nuclide. The more you've got, the toastier it is. This is a simple yet powerful equation that relates amount of material with activity. If the number of atoms is known, it can be converted to moles of atoms using Avogadro's number (6.022×10^{23} atoms/mol), and then into mass using the atomic mass (g/mol). Likewise, if the mass is known, the number of atoms can be calculated, as well as activity.

Example: What is the mass of $1.00\ \mu Ci$ of ^{137}Cs?

First convert activity to dpm:

$$1.00\,\mu\,Ci \times \frac{2.22 \times 10^{6}\ dpm}{\mu\,Ci} = 2.22 \times 10^{6}\ dpm$$

Next, calculate the number of ^{137}Cs atoms (after looking up $t_{1/2}$ for ^{137}Cs):

$$A = kN, \quad N = \frac{A}{k} = \frac{A \cdot t_{1/2}}{\ln 2} = \frac{(2.22 \times 10^{6}\ dpm)(30.07\,a)}{\ln 2}$$

But wait! Notice the problem with the time units? Decays per minute doesn't cancel out when multiplied by years. To fix this, it's handy to know that there are 5.256×10^{5} min in a year:

$$N = \frac{(2.22 \times 10^{6}\ dpm) \times (30.07\ a) \times [(5.256 \times 10^{5}\ min)/a]}{\ln 2} = 5.06 \times 10^{13}\ atoms$$

Convert to mass:

$$(5.06 \times 10^{13}\ atoms) \times \frac{mol}{6.022 \times 10^{23}\ atoms} \times \frac{137\,g}{mol} = 1.15 \times 10^{-8}\,g$$

Wow! That's a very small mass for something spitting out radiation two million times each minute! A little radioactive material can often go a long way. Notice also that 137 g/mol was used as the atomic mass. The mass number of a particular nuclide can always be used as an approximate atomic mass. As we'll see later in this chapter, more precise atomic masses can be calculated (or looked up), but an approximate value is adequate for the problem above.

Because of the direct proportionality of A and N, the decay equations given at the beginning of this chapter can be rewritten as:

$$\ln \frac{N_2}{N_1} = -kt \qquad (2.6)$$

$$N_2 = N_1\, e^{-kt} \qquad (2.7)$$

This is particularly handy if we know the mass or number of a radioactive nuclide.

2.3   SPECIFIC ACTIVITY

Activity can easily be measured using a detector with a known efficiency. Mass is also easily measured using an appropriate balance. So why is the equation $A = kN$ a big deal? Even if atoms of a particular nuclide are painstakingly purified, it is not possible to exactly determine the mass of the nuclides that have not yet decayed. In other words, there will always be some contribution to the mass by the product nuclide(s). Additionally, radioactive nuclides are often incorporated into molecules or other nonradioactive materials. If mass is determined, it is the mass of all atoms present, not just the radioactive ones. Therefore, determining mass (and thereby N) of a radioactive nuclide is not as simple as throwing it onto a balance.

Occasionally, it is important to know how radioactive a sample is per gram of sample, regardless of what is in it. **Specific activity** is the activity per unit mass of a substance, whether it is atoms of a single nuclide, a compound, or some mixture of compounds. The units are typically Ci/g. Let's start simple and calculate the specific activity of a pure nuclide.

Example: Calculate the specific activity of ^{133}Xe.

Notice that the problem doesn't give us a mass! The specific activity will be the same no matter how much ^{133}Xe there is. Assume exactly 1 g, then calculate N:

$$1\,g \times \frac{\text{mol}}{133\,g} \times \frac{6.022 \times 10^{23}\ \text{atoms}}{\text{mol}} = 4.53 \times 10^{21}\ \text{atoms}$$

Now calculate activity:

$$A = kN = \frac{\ln 2}{5.243\ \text{d}} \times (4.53 \times 10^{21}\ \text{atoms}) = 5.99 \times 10^{20}\ \text{decays per day}$$

Specific activity is:

$$\frac{\dfrac{5.99 \times 10^{20}\ \text{decays}}{\text{day}} \times \dfrac{\text{d}}{24\,\text{h}} \times \dfrac{\text{h}}{60\,\text{min}} \times \dfrac{\text{min}}{60\,\text{s}} \times \dfrac{\text{Ci}}{3.7 \times 10^{10}\ \text{dps}}}{1\,g} = 1.87 \times 10^{5}\ \frac{\text{Ci}}{g}$$

Example: Calculate the specific activity of a gas mixture containing 78 g N_2, 21 g O_2, and 1.00 g of ^{133}Xe.

There are no naturally occurring radioactive isotopes of nitrogen or of oxygen; therefore ^{133}Xe is the only hot component of this mixture. Its activity is:

$$\frac{1.87 \times 10^{5}\,\text{Ci}}{g} \times 1.00\,g = 1.87 \times 10^{5}\,\text{Ci}$$

Ok, maybe that was trivial, but this *is* only Chapter 2 ... the final step is to divide by the total mass of the sample:

$$\frac{1.87 \times 10^5 \, \text{Ci}}{78\,g + 21\,g + 1.00\,g} = 1.87 \times 10^3 \, \frac{\text{Ci}}{g}$$

Notice that specific activity decreases as a radioactive nuclide becomes more "dilute" within a sample.

2.4 DATING

The fact that radioactive decay can easily be predicted through mathematics opens up some interesting practical applications. One application is **radioactive dating**—the determination of the age of an object by measuring how much radioactive material is left in it. The most common form of dating is carbon dating. Small amounts of radioactive ^{14}C are continuously produced in the upper atmosphere through the bombardment of cosmic rays (Chapter 7). ^{14}C has a half-life of 5715 years, so it is also continually decaying. The rates of production have long since balanced out with the rate of decay, meaning that there has been a small but constant amount of ^{14}C on the planet for almost all of its history. Carbon is an essential element for all living things; therefore a small fraction of their carbon is ^{14}C. Once an organism dies, it no longer takes in ^{14}C from the biosphere, and the ^{14}C in the dead thing slowly begins to decay. Over time the ratio of ^{14}C to nonradioactive carbon (mostly ^{12}C) gradually decreases.

The specific activity of carbon is a convenient way to express the relative amount of ^{14}C present in all carbon. Prior to 1950, the specific activity of carbon on earth was roughly 14 dpm/g. It varied somewhat before that time because cosmic ray bombardment has not always been constant. The specific activity of carbon has varied in recent decades, primarily because of the tremendous amount of fossil fuels being burned. Fossil fuels are made up of a lot of carbon from living things that died quite a long time ago. Since the time that they died, all of the ^{14}C that was once part of them has decayed away. Burning fossil fuels therefore pumps the atmosphere full of nonradioactive isotopes of carbon, lowering the specific activity of all carbon. It causes other problems too, but they are topics for other books.

Carbon dating works well for dating stuff that has been dead less than ~60,000 years (~10 half-lives). If it's been dead longer than that, it may be difficult to accurately measure the amount of ^{14}C present. Carbon dating was famously used to date the Shroud of Turin, which is believed by some to be the burial shroud of Jesus.

Example: The specific activity of carbon from the Shroud of Turin is 12.8 dpm/g today. When were the plants killed to make this cloth?

$$\ln \frac{A_2}{A_1} = -\left(\frac{\ln 2}{t_{1/2}}\right)t \quad t = \frac{\ln \frac{A_2}{A_1}}{-\frac{\ln 2}{t_{1/2}}} = \frac{\ln \frac{12.8}{14}}{-\frac{\ln 2}{5715\,a}} = 740 \text{ a}$$

Therefore the cloth was made 740 years ago, or approximately 1300 AD.

Other nuclides can be used for similar dating methods. For example, the age of rocks containing ^{238}U can be determined by measuring the ^{238}U:^{206}Pb ratio, assuming that all of the ^{206}Pb now present in the rock comes from the ^{238}U decay series. The same can be done when ^{87}Rb is found in a rock.

Example: A meteorite is found to contain 0.530 g of ^{87}Rb and 0.042 g of ^{87}Sr. If 5 mg of the ^{87}Sr was present when the rock was formed, how old is the rock?

Over time, ^{87}Rb undergoes beta decay to ^{87}Sr:

$$^{87}_{37}\text{Rb} \rightarrow \,^{87}_{38}\text{Sr} + \,^{0}_{-1}e$$

How much ^{87}Rb was present in the rock when it was formed? It should be the sum of what's there now, and what decayed to ^{87}Sr:

$$0.530\,\text{g }^{87}\text{Rb} + \left((0.042\,\text{g }^{87}\text{Sr} - 0.005\,\text{g }^{87}\text{Sr}) \times \frac{\text{mol }^{87}\text{Sr}}{87\,\text{g }^{87}\text{Sr}} \times \frac{1\,\text{mol }^{87}\text{Rb}}{1\,\text{mol }^{87}\text{Sr}} \times \frac{87\,\text{g }^{87}\text{Rb}}{\text{mol }^{87}\text{Rb}} \right)$$

$$= 0.567\text{g }^{87}\text{Rb}$$

Notice that the conversion from grams of ^{87}Sr to grams of ^{87}Rb is really unnecessary since they have the same mass and one ^{87}Sr is produced for every ^{87}Rb that decays. This is often the case in dating problems, but not always.

Now we can use the same formula as the ^{14}C example, except here we use the gram ratio instead of the activity ratio (since $A \propto N$ and N is proportional to mass, so long as the nuclide is the same).

$$t = \frac{\ln \dfrac{N_2}{N_1}}{-\dfrac{\ln 2}{t_{1/2}}} = \frac{\ln \left(\dfrac{0.530\,\text{g}}{0.567\,\text{g}} \right)}{-\dfrac{\ln 2}{4.88 \times 10^{10}\,\text{a}}} = 4.75 \times 10^9 \,\text{a}$$

2.5 BRANCHED DECAY

Radioactive decay is often a messy process. As we've already seen in the naturally occurring decay series, some nuclides make matters worse by decaying by more than one pathway. ^{64}Cu is a classic example of a nuclide that has more than one decay mode. It can decay by β^- to ^{64}Zn, or β^+ or electron capture to ^{64}Ni. The half-life for ^{64}Cu is given as 12.701 h, but this accounts for all decays. Information on the relative percentages of each form of decay, or **branch ratio**, is given in table 2.2. These data are from the Korean Atomic Energy Research Institute (http://atom.kaeri.re.kr/ton/).

Table 2.2 tells us that ^{64}Cu decays via β^- only 39% of the time, and the remaining 61% of the time it decays via positron emission or electron capture. Notice that the percentages in table 2.1 don't quite add up to 100%, because they're real-world data!

TABLE 2.2

Branch Ratios for ^{64}Cu

β⁻	39.00%	^{64}Zn
β⁺	17.40%	^{64}Ni
EC	43.57%	^{64}Ni

We can experimentally determine half-life a couple of different ways. We can observe the change in count rate over time and use

$$t_{1/2} = -\frac{\ln 2 \times t}{\ln(A_2/A_1)}$$

which is a simple rearrangement of equation (2.1). Using this method does not require application of the branch ratio—the change in activity (or count rate) with time does not depend on the branch ratio.

The other way to find half-life is to determine the activity and mass of the nuclide (convert to N) and use:

$$t_{1/2} = \frac{\ln 2 \times N}{A}$$

which is a rearrangement of $A = kN$. As you might guess, the first method is more useful when the nuclide has a short half-life and the second method works best for nuclides with long half-lives. If this latter method is used with a nuclide that undergoes branched decay, the branch ratio must be taken into account to properly calculate activity. Activity results from all decays. If only one mode is observed during an experiment, branch ratio and efficiency must be used to calculate activity. It's a bit like trying to figure out how quickly a bucket of water will drain through a hole based on the amount of water in the bucket and the flow rate out the hole. The problem is that the bucket has another hole that you can't see. The bucket will always drain faster than you think it should.

Example: A sample of copper is known to contain 9.2×10^7 atoms of ^{64}Cu, and registers 2345 cpm on a detector that can only detect beta particles (β⁻). If this detector has an efficiency of 7.20%, what is the half-life of ^{64}Cu?

First calculate dpm using the percent efficiency:

$$\text{dpm} = \frac{2345\,\text{cpm}}{7.2\%} \times 100\% = 3.3 \times 10^4\,\text{dpm}$$

Next, account for the branch ratio. We know the decay rate above is only for β⁻, so the decay rate for all decays must be higher, so we divide by 0.3900.

$$\frac{3.3 \times 10^4\,\text{dpm}}{0.3900} = 8.4 \times 10^4\,\text{dpm}$$

Now plug into $A = kN$:

$$t_{1/2} = \frac{\ln 2 \times (9.2 \times 10^7 \text{ atoms})}{8.4 \times 10^4 \text{ dpm}} = 760 \text{ min}$$

Because the units for activity were dpm, the units on half-life are minutes. It looks like this half-life is more conveniently expressed in hours:

$$t_{1/2} = 760 \text{ min} \times \frac{1 \text{ h}}{60 \text{ min}} = 13 \text{ h}$$

With only two significant figures to work with, this is pretty close to the actual value of 12.701 h.

If we had left out the branch ratio in the above calculations, the half-life would be 33 h—quite a ways off! A half-life calculated using data from one branch of the decay is called a **partial half-life**. A partial half-life is always longer than the true or total half-life. If we know the fraction of the decays that go by each branch, we can calculate the partial half-life by simply dividing the total half-life by the fraction. Since the true half-life for ^{64}Cu is 12.701 h, the electron capture (EC) partial half-life is:

$$\frac{12.701 \text{ h}}{0.4357} = 29.15 \text{ h}$$

This makes sense! Since only about half of the decays are EC, the half-life should appear to be about twice as long.

Example: ^{192}Ir decays via β^- emission 95.13% of the time. If its β^- partial half-life is observed to be 77.6 d, what is its true half-life?

$$77.6 \text{ d} \times 0.9513 = 73.8 \text{ d}$$

2.6 EQUILIBRIA

The naturally occurring decay series that begins with ^{238}U starts with an alpha decay forming ^{234}Th. ^{234}Th is also unstable toward decay, spitting out a beta particle, forming ^{234}Pa. These two decay processes can be combined into a single equation:

$$^{238}_{92}\text{U} \xrightarrow{\alpha} {}^{234}_{90}\text{Th} \xrightarrow{\beta} {}^{234}_{91}\text{Pa}$$

In any decay, the original nuclide is called the **parent** nuclide, and the product nuclide is called the **daughter**. In the case of a decay series, all nuclides formed after the parent decays are called daughters. In the example above, ^{238}U is the parent and ^{234}Th and ^{234}Pa are the daughters. A decay series could be generically represented by:

$$A \xrightarrow{t_{1/2(A)}} B \xrightarrow{t_{1/2(B)}} C \ldots$$

Since daughter B is constantly being formed by decay of parent A, and is constantly decaying to daughter C, how can we figure out how much B exists at any one moment in time? More generally, how can we determine the total activity of the sample or the amount of any one nuclide at a particular point in time? It seems almost too complex to comprehend, but we can look at three simple scenarios where the math becomes more accessible.

2.6.1 SECULAR EQUILIBRIUM

If the half-life of the parent is very long relative to the half-life of the daughter, and the parent is isolated at time $= 0$, then the following equations describe the system *at any time* after the parent is isolated:

$$A_B = A_A(1 - e^{-k_B t}) \tag{2.8}$$

$$N_B = \frac{k_A}{k_B} N_A (1 - e^{-k_B t}) \tag{2.9}$$

The subscripts A and B refer to parent and daughter, respectively. Therefore, A_B is the activity of the daughter and k_A is the decay constant for the parent. These two equations are identical. To derive the second from the first, substitute $k_X N_X$ ($X = A, B$) for the activities.

If the activities of both parent and daughter are plotted after the isolation of pure parent, something like figure 2.2 will result. Because the parent half-life is so incredibly long, its activity is constant over the period of observation. Daughter activity starts at zero because the parent was chemically isolated. As daughter is produced,

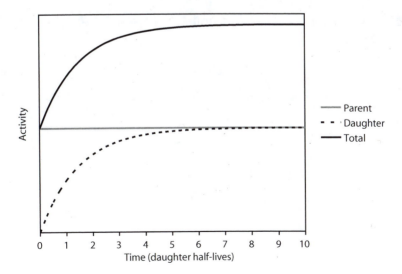

FIGURE 2.2 Secular equilibrium.

its activity gradually increases. However, as the amount of daughter builds up, its rate of decay gradually becomes more significant. When the rate of formation of B is equal to the rate of decay of B, it has reached **secular equilibrium**, much like the ^{14}C in our biosphere discussed in the Dating section of this chapter. Secular equilibrium is reached roughly seven daughter half-lives after the parent is isolated. At this point the activity of the daughter is equal to the activity of the parent, and the activity equation given above can be simplified to $A_B = A_A$. This is true for all radioactive daughters (not just B) of a long-lived parent A. If C is a stable nuclide, the total activity after secular equilibrium is reached is twice that of either the parent or the daughter, as can be seen in figure 2.2.

Because $A_B = A_A$ at any point in time *after* secular equilibrium is reached, we can substitute in kN for the activities, and generate more equations like:

$$k_A N_A = k_B N_B \tag{2.10}$$

$$\frac{N_B}{N_A} = \frac{k_A}{k_B} = \frac{t_{1/2(B)}}{t_{1/2(A)}} \tag{2.11}$$

These equations can only be applied after secular equilibrium is reached. Secular equilibrium is very handy for solving all kinds of problems. It is especially useful with the three naturally occurring radioactive decay series. As you may recall, these decay series are a significant source of many of the radionuclides found in nature. Since they all start with a long-lived nuclide, and have been around for a few billion years, secular equilibrium applies!

1. Since the activities of all nuclides in secular equilibrium are equal, the mass ratios can easily be determined.
2. If the parent has a very long half-life, it is difficult to determine accurately by sitting around waiting to observe a change in parental activity over time. If we know the half-life of one of the daughters, and its mass ratio to the parent, we can calculate the half-life of the long-lived parent using:

$$\frac{N_B}{N_A} = \frac{t_{1/2(B)}}{t_{1/2(A)}}$$

where the mass ratio between daughter and parent is N_B/N_A.
3. The mass of the parent can be determined from the activity of any daughter.
4. The mass of the parent can be used to calculate the activity of a daughter.

Example: 10.2 kg of U was isolated from some ore, which was also found to contain 3.33×10^{-3} g of ^{231}Pa. What is the half-life of ^{235}U?

$$t_{1/2(\text{U-235})} = \frac{N_U}{N_{Pa}} t_{1/2(\text{Pa-231})}$$

$$= \frac{10{,}100\ \text{g} \times \dfrac{\text{mol U}}{238.03\ \text{g}} \times \dfrac{6.022 \times 10^{23}\ \text{atoms}}{\text{mol}} \times \dfrac{0.7200\ \text{atoms}\ ^{235}\text{U}}{100\ \text{atoms U}}}{\left(3.33 \times 10^{-3}\ \text{g}\right) \times \dfrac{\text{mol}\ ^{231}\text{Pa}}{231.04\ \text{g}} \times \dfrac{6.022 \times 10^{23}\ \text{atoms}}{\text{mol}}} \times (3.28 \times 10^4\ \text{a})$$

$$= 7.02 \times 10^8\ \text{a}$$

It is important to realize that uranium isolated from ore will contain all naturally occurring isotopes of uranium in their normal proportions. To solve the problem above, the number of ^{235}U atoms within 10.1 kg of uranium must be calculated. Notice that the conversion from moles to atoms (Avogadro's number) is unnecessary, as it cancels out.

Example: What is the activity of ^{222}Rn in an ore sample containing 3.00 g of uranium?

$$A_{\text{Rn-222}} = A_{\text{U-238}} = k_{\text{U-238}} N_{\text{U-238}}$$

$$= \left(\frac{\ln 2}{4.47 \times 10^9\ \text{a}}\right)\left(3.00\ \text{g U} \times \frac{\text{mol U}}{238.03\ \text{g U}} \times \frac{6.022 \times 10^{23}\ \text{atoms U}}{\text{mol U}} \times \frac{99.2745\ \text{atoms}\ ^{238}\text{U}}{100\ \text{atoms U}}\right)$$

$$= 1.17 \times 10^{12}\ \text{decays per year}$$

$$= 2.23 \times 10^6\ \text{dpm}$$

In this example, the correct decay series must first be determined. ^{222}Rn is only observed in the ^{238}U decay series; therefore it is only produced by ^{238}U.

The above discussion of secular equilibria assumes that the parent does not undergo branched decay and that no branched decay happens between parent and daughter that doesn't also go through the daughter. Close examination of the nuclides involved in the two examples given above should bring a sigh of relief. No branches. If the parent (A) undergoes branched decay, then the activity of the daughter (A_B) is equal to the activity of the parent (A_A) times the branch ratio (BR):

$$A_B = A_A \times BR \tag{2.12}$$

2.6.2 Transient Equilibrium

Transient equilibrium is like secular equilibrium in that the half-life of the parent is longer than that of the daughter. The difference is the time of observation. An equilibrium is considered transient if significant parental decay occurs during the time of observation (usually more than 7–10 daughter half-lives). In a transient equilibrium the parent and daughter half-lives tend to be closer together in value.

If the parent is isolated at $t = 0$, then as the daughter is produced and begins to decay, the parent is also decaying. Parental activity will decrease in an exponential

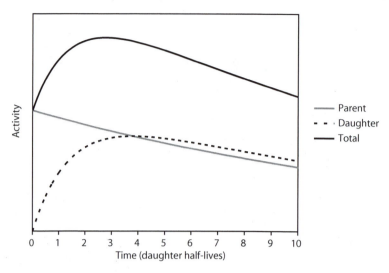

FIGURE 2.3   Transient equilibrium.

fashion, as shown in figure 2.3. Daughter activity grows at first, peaks, then decreases, eventually paralleling the decrease in parent activity. Notice that daughter activity eventually exceeds parent activity. It is after ~7 daughter half-lives that the relative activities of parent and daughter stabilize, establishing transient equilibrium. Total activity again starts at the same point as parent activity, increases as the amount of daughter increases, and then parallels the decrease in parent and daughter activity. *After* equilibrium is achieved, the following equations can be applied:

$$\frac{A_A}{A_B} = 1 - \frac{t_{1/2(B)}}{t_{1/2(A)}} \tag{2.13}$$

$$\frac{N_B}{N_A} = \frac{t_{1/2(B)}}{t_{1/2(A)} - t_{1/2(B)}} \tag{2.14}$$

Example: A sample of ^{47}Ca is isolated from other radionuclides. Thirty-one days later the activity due to ^{47}Ca is 2.19×10^4 dpm. What activity would you expect for ^{47}Sc?

The half-life for ^{47}Ca is 4.536 d and for ^{47}Sc it is 3.349 d. This sample has attained transient equilibrium.

$$A_{Sc-47} = \frac{A_{Ca-47}}{1 - \dfrac{t_{1/2(Sc-47)}}{t_{1/2(Ca-47)}}} = \frac{2.19 \times 10^4 \text{ dpm}}{1 - \dfrac{3.349 \text{ d}}{4.536 \text{ d}}} = 8.37 \times 10^4 \text{ dpm}$$

For all transient equilibria, daughter activity is maximized at time t_{max}, which can be calculated from the parent and daughter half-lives:

$$t_{max} = \left[\frac{1.44 t_{1/2(A)} t_{1/2(B)}}{\left(t_{1/2(A)} - t_{1/2(B)}\right)}\right] \times \ln\frac{t_{1/2(A)}}{t_{1/2(B)}} \tag{2.15}$$

Example: In the previous example, how long did it take the daughter to reach maximum activity?

$$t_{max} = \left[\frac{1.44 t_{1/2(Ca\text{-}47)} t_{1/2(Sc\text{-}47)}}{\left(t_{1/2(Ca\text{-}47)} - t_{1/2(Sc\text{-}47)}\right)}\right] \times \ln\frac{t_{1/2(Ca\text{-}47)}}{t_{1/2(Sc\text{-}47)}}$$

$$= \left[\frac{1.44 \times 4.536\,d \times 3.349\,d}{\left(4.536\,d - 3.349\,d\right)}\right] \times \ln\left(\frac{4.536\,d}{3.349\,d}\right)$$

$$= 5.59\,d$$

2.6.3 No Equilibrium

When the half-life of the parent is shorter than that of the daughter, no equilibrium is reached. If pure parent is again separated at time $t=0$, the parent activity will decrease rapidly, as shown in figure 2.4. Daughter activity will increase until the parent is nearing depletion, then it will gradually decrease. When the parent is gone (>7 parent half-lives), the daughter's decrease in activity will be based entirely on its own half-life ($t_{1/2(B)}$). Since no equilibrium is reached, no special equations

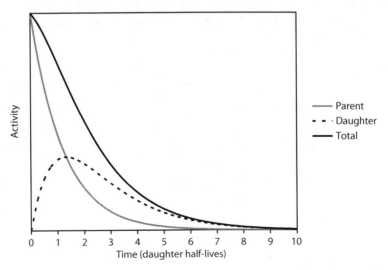

FIGURE 2.4 Equilibrium is not established when $t_{1/2(A)} < t_{1/2(B)}$.

are necessary to describe the change in daughter activity after the parent is gone. The activity equations given at the beginning of the chapter are sufficient. However, equation (2.15), used to calculate maximum daughter activity for transient equilibria, also applies here:

$$t_{max} = \left[\frac{1.44 t_{1/2(A)} t_{1/2(B)}}{\left(t_{1/2(A)} - t_{1/2(B)} \right)} \right] \times \ln \frac{t_{1/2(A)}}{t_{1/2(B)}}$$

Nuclear equilibria are sometimes confusing to students. Perhaps it's because the term 'equilibrium' is not the best one to describe what is actually happening. We say the nuclear equilibria are reached when the relative rate of formation of daughter equals its relative rate of decay ($A \rightarrow B \rightarrow C...$). The more familiar use of the word 'equilibrium' likely comes from chemistry. In a chemical reaction, equilibrium is reached when the rate of formation of products equals the rate of formation of reactants. In other words, reactants are combining to form products at the same rate that products are combining to form reactants ($A \leftrightarrows B$). For a radioactive decay, this would mean that the daughter would have to "decay" back into the parent, which is not possible. "Steady-state" would be a better term to describe nuclear equilibria. Steady-state is like pouring water into a bucket that's already full—the rate water is being added is equal to the rate it is spilling out, but we always have the same amount of water in the bucket.

Two other points of confusion are trying to decide if the system under investigation is at equilibrium, and what kind of equilibrium it is. In this text, the passage of seven daughter half-lives is consistently used as the point where equilibrium is first reached. Other texts state that equilibrium is only attained after 10 daughter half-lives have elapsed. If the point in time is within the controversial $7–10 \times t_{1/2(B)}$ period, then it is largely a matter of the level of precision required. For most work, seven half-lives will be sufficient, and that's why it is used as a point of demarcation in this book. However, if very precise measurements are being made, then waiting until 10 half-lives have passed before declaring the system to be at equilibrium may be necessary.

The confusion in determining whether secular or transient equilibrium is appropriate to a particular problem can also be ambiguous. The half-life of the parent needs to be long compared to the time of observation for the equilibrium to be secular. In other words, the parent cannot decay significantly ($<5–10\%$) during the time of observation. Again, the exact percentage of decay that is tolerable will depend on the level of precision required by the problem.

2.7 STATISTICS

Radioactive decay is a random event. We can't say when an unstable nucleus will decay, only that it will probably decay within seven or so half-lives. If we've got a whole lot of an unstable nuclide, the number of them that decay in a certain time will vary, even if the number of these nuclides does not change significantly. Table 2.3 gives count data for a ^{204}Tl source counted 20 times for one minute each. With a

---

TABLE 2.3
Count Data

Run	cpm
1	2123
2	2189
3	2120
4	2167
5	2125
6	2217
7	2192
8	2098
9	2154
10	2130
11	2112
12	2151
13	2191
14	2124
15	2071
16	2085
17	2179
18	2134
19	2096
20	2187

---

half-life of 3.8 a, it did not decay appreciably during the 20 min these data were collected. The lowest count is 2071 cpm and the highest 2217 cpm—a difference of 146 cpm! So what is the true number of cpm for this sample?

As you might guess, the best we can do is average our data and hope it is close to the true value. How much error is in this estimate? Statistics can give us the answers!

The probability that we will get a particular count rate can be approximated by a Gaussian (or normal) distribution. A Gaussian is a curve that is symmetric about the mean (true) value, as shown in figure 2.5. The vertical axis of the plot corresponds to the probability that a value along the horizontal axis will be observed. The curve is highest at the mean and tapers away in both directions, so we are much more likely to get a count rate close to the mean than far from it.

The uncertainty associated with our measurement can be approximated by standard deviation, which is the square root of the mean value:

$$\sigma \approx \sqrt{\overline{x}} \tag{2.16}$$

This equation applies no matter how many measurements are made, even if there is only one. This calculation of standard deviation is likely a little different than

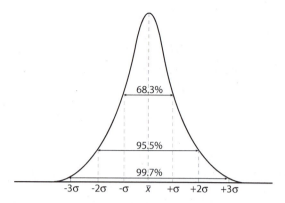

FIGURE 2.5 A Gaussian distribution.

the one on your calculator or in Microsoft® Excel, as it applies only to statistically random processes (such as radioactive decay). Most nuclear scientists use uncertainties expressed to 2σ, which means that 95.5% of the time an additional count will fall within this range. For our sample data (table 2.3) we could state that the count mean is 2142 ± 92 cpm to a confidence level of 2σ. The 2σ range of these counts is 2050–2235 cpm. In other words, if we collected data for another one-minute count, 95.5% of the time it will fall between 2050 and 2235 cpm. Note that none of the 20 counts in table 2.3 exceed these limits.

The mean and standard deviation allow us to look at a single set of data and characterize their precision. A relatively large standard deviation means poor precision and a relatively low σ means good precision. A good way to compare the precision for two or more sets of data is to calculate the relative standard deviation (RSD). RSD is determined by dividing the standard deviation by the mean value:

$$RSD = \frac{\sigma}{\bar{x}} \tag{2.17}$$

The RSD for the data in table 2.3 is 0.022.

Sometimes we would like to compare the average count of one set of data points with the average count from another set of data points. If we were to repeat the above experiment (table 2.3) and obtain a new set of 20 one-minute counts, the mean of the new set (population) could be compared with the mean of the first population using the standard deviation of the mean (σ_{mean}). It is obtained by dividing the standard deviation by N:

$$\sigma_{mean} = \frac{\sigma}{\sqrt{N}} \tag{2.18}$$

where N is the number of data (runs) in the population. For the data given in table 2.3:

$$\sigma_{mean} = \frac{46}{\sqrt{20}} = 10 \text{ cpm}$$

This is 20 cpm at the $2\sigma_{mean}$ confidence level. We could then quote the mean and its uncertainty as 2142 ± 20 cpm. This suggests that the new mean, calculated from the second set of data, should fall in the range of 2122–2162 cpm. You should note that the uncertainty will become smaller as the number of runs increases.

Experimentally, the mean obtained in a series of 20 one-minute counts should have the same uncertainty as that obtained from one 20-min count. To calculate σ_{mean} for a single count, we can simply replace the number of trials (N) in the equation for σ_{mean} with the time, in minutes, for the single count.

As an example, let's say the total counts for the lone 20-min count was 43,000 counts. The average or mean count is obtained (as before) by dividing the total counts by the total time of the counts. The average count rate ($\bar{x}$) is calculated as 43,000 counts/20 min = 2150 cpm. The standard deviation is calculated as before:

$$\sigma \approx \sqrt{\bar{x}} = 46$$

The count mean is 2150 ± 92 cpm to a confidence level of 2σ. In other words, if we collected an additional one-minute count, we would expect it to be somewhere between 2058 and 2242 cpm; pretty much the same result as with the 20 one-minute counts set of data.

What we would like to do now is compare the σ_{mean} values for the two sets of data we have collected. The standard deviation of a mean obtained for a count longer than one minute is:

$$\sigma_{mean} = \frac{\sigma}{\sqrt{t}} \tag{2.19}$$

where t is time in minutes. For our second data set the standard deviation of the mean is:

$$\sigma_{mean} = \frac{46}{\sqrt{20}} = 10$$

Therefore, the mean of the 20-min count is 2150 ± 20 cpm to a certainty of 95.5%. This is a range of 2130–2170 cpm. If we now look at the mean ranges for the two data sets, we have 2122–2162 cpm and 2130–2170 cpm. Since these ranges overlap, the two mean values are "not significantly different from each other." Although the numerical values of the mean values are different, the difference is not large enough to be statistically significant. Many more runs or a much longer counting time (lowering σ) would be needed to determine if the mean values are, in fact, different.

If we had a very large sample of data we could calculate the true mean (usually called μ). The standard deviation (σ) would then be exactly equal to $\sqrt{\mu}$. In a limited number of counts, like the 20 counts in our first data set, we can calculate a good value of σ by using the average of our counts ($\bar{x}$) instead of the true mean μ. But we should keep in mind that this is an approximation because $\bar{x}$ is only an approximation of the true mean μ.

Readers should take care not to confuse the meanings of σ and σ_{mean}. Both are expressions of uncertainty and reflect on the precision of the data. Standard deviation (σ), establishes the range within which we can expect another *single* value to appear. For the data in table 2.3, it tells us that there's a 95.5% chance that another one-minute count run will fall between 2050 and 2235 cpm. The standard deviation of the mean (σ_{mean}) establishes the range within which we can expect another *set* of runs to produce a mean value. For the data in table 2.3, it tells us that there's a 95.5% chance that the mean value of another set of one-minute counts will fall between 2122 and 2162 cpm.

We can still calculate an uncertainty for a single count x, assuming it is a reasonable approximation of the average of many counts $\bar{x}$. The uncertainty of a single count is found by $\sigma \approx \sqrt{x}$. For a confidence level of 95.5% or 2σ, this is $2\sqrt{x}$. That is, two times the square root of a count is the uncertainty of each count. Please note that σ must be calculated using the raw count data. For example, if a source exhibits 2345 counts between $t = 120$ and $140\,s$, then $2\sigma = 2 \times \sqrt{2345} = 97$. When the count rate (cpm) is calculated, 2σ is also multiplied by 3. Other (such as coincidence and background) corrections are applied to the count rate but not 2σ. The final count rate and random error for this example are 7068 ± 291 cpm, or about a 4% uncertainty.

How can this be expressed directly on a plot? Since count data are typically plotted as ln cpm, we need to find the uncertainty range in these units. In the example given above, the highest count rate in the 2σ range is $7068 + 29 = 7359$ cpm. Since $\ln 7068 = 8.86$ and $\ln 7359 = 8.90$, our 2σ range on the plot would be 8.86 ± 0.04. If 2σ ranges are calculated on an Excel spreadsheet for all count data, they can be incorporated into the graph by right clicking a data point, selecting Format Data Series, then the Y Error Bars tab. Select Both under Display, and click next to Custom under Error Amount, then input the appropriate cell ranges next to $+$ and $-$. In plotting these data, the count would be represented by a point and the range by a vertical bar extending above and below the point. In reality, the point could've fallen anywhere along that line. An example plot, using the decay of a ^{108}Ag sample, is shown in figure 2.6.

From this statistical analysis you can see that higher count data (longer count times or a higher activity sample) will have a smaller uncertainty and should be favored in drawing the best-fit line to experimental data. This is clearly evident in figure 2.6—as ln cpm decreases, the error bars get longer. The computer takes each point as equally valid and calculates the best line. You may find that you get better results if you eliminate some points from the trend line. Points at the lowest count rates can be eliminated because of the higher uncertainty associated with them. Points that are more than $\pm 2\sigma$ from the expected count may be viewed with skepticism knowing that only 4.5% of all points are expected to fall outside of this range (100%−95.5% = 4.5%). Points that are beyond 3σ of the expected value (a 99.7% confidence level) *could* be eliminated from most data as they are probably due to mistakes made in collecting the data. Students are cautioned against simply discarding data because they are outside some level of uncertainty. Unless you have some other reason to suspect there is a problem with a particular data point, it should be retained.

As stated above, radioactive decay data closely approximate a Gaussian distribution. In reality, radioactive data are better fit to the similar, but mathematically more

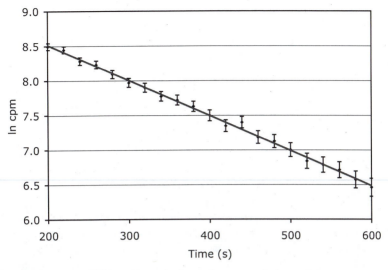

FIGURE 2.6 Decay of [108]Ag with vertical error bars on each data point.

complex, Poisson distribution. To determine whether a set of data follows a Poisson distribution, we need to calculate χ^2 (chi-squared):

$$\chi^2 = \frac{1}{\overline{x}}\sum\left(\overline{x} - x\right)^2 \qquad (2.20)$$

To find χ^2, we'll need to find the square of the difference of every individual value in the data set with the mean value $((\overline{x} - x)^2)$, add them up $(\Sigma(\overline{x} - x)^2)$, and then multiply by the reciprocal of the mean $(1/\overline{x})$. Sounds crazy, but it works. Try calculating this value using the data in table 2.3; χ^2 should equal 14.9.

Is this good? To interpret a chi-squared value, we'll need to use a table of chi-squared like the one found in Appendix C. To use this table, you need to know that there are 20 data points; therefore the data have 19 degrees of freedom $(N-1)$. Reading down the first column in to 19, there are several values of chi-squared listed in the body of the table to be compared with the calculated value. For 19 degrees of freedom there is 1 chance in 100 that chi-squared will be smaller than 7.633 and also 1 chance in 100 that it will be larger than 36.19. Our value of 14.9 certainly falls within that range.

A more reasonable set of limits is the 80% chance. There is a 10% chance that chi-squared will be smaller than 11.65 for 19 degrees of freedom. This is obtained by reading down the 0.90 column to 19 degrees of freedom. There is also a 10% chance that chi-squared will be larger than 27.20. This is obtained by reading down the 0.10 column to 19 degrees of freedom. Therefore, there is a 80% chance that chi-squared will fall between 11.65 and 27.20 for a Poisson distribution. A 80% chance means that 8 out of 10 times this will happen. Once again, our value of 14.9 fits within the range, giving us confidence that our data fit a Poisson distribution (as well they should!).

There have probably been many times in your life when you did not have a chi-squared table at hand. If you look down the 0.50 column, you will note that chi-squared increases with the number of trials N. The value of the quantity chi-squared is roughly $N-2$. If you ever need a ballpark estimate of chi-squared for, say, 30 determinations, it would be about 28. The entry in the table for 29 degrees of freedom, that is 30 points at the 50% probability, is 28.34. Darn close.

So far, we've only discussed the inherently random nature of radioactive decay, which leads to a certain amount of uncertainty in data collected from radioactive sources. This is truly random error, but the wonderful thing about it is that we can always calculate it. What if we carefully calculate the random error associated with some decay measurements, but it is clear some additional error is also present. That error would be systematic error, and is likely due to some external change during the experiment. Perhaps the source or the detector moved, or some shielding between the source and detector was added or removed. It could be a number of factors. A good scientist will take great care in making measurements and consider many possibilities if systematic error is apparent.

QUESTIONS

1. Calculate the average atomic mass of naturally occurring Mg.
2. Naturally occurring rubidium is a mixture of only two isotopes: ^{85}Rb (84.9118 u) and ^{87}Rb (86.9092 u). If the average atomic mass for Rb is 85.4678 u, calculate the percent abundance of its two isotopes.
3. What is the difference between count rate and decay rate? How are they related, mathematically?
4. What fraction of the original activity is left after 3 half-lives have passed? After 5?
5. Counts from a radionuclide decrease from 5718 to 515 dpm in 24.0 h. What is the half-life of this nuclide?
6. A sample of ^{99}Mo has an activity of 15,000 dpm at noon today. Calculate its activity 30.0 days from now.
7. Using a ^{204}Tl source that was calibrated at 1.0 μCi on September 3, 1993, you observe 1600, 939, and 539 cpm on the first, second, and third shelves below your detector. Calculate the percent efficiency on all three shelves.
8. Calculate the activity due to the decay of ^{40}K in your body. It will help to know that there are 140 g of potassium in a person weighing 70 kg.
9. Calculate the specific activity of 0.353 g of NaTcO$_4$. Assume all of the technetium is ^{99}Tc, and it is the only radioactive nuclide present.
10. Calculate the specific activity (μCi/g) of naturally occurring platinum.
11. A piece of wood taken from an Egyptian tomb shows 55.9% of the ^{14}C activity that a living piece of wood did during the approximate time the tomb was constructed. When did the tree containing this piece of wood die?
12. A rock is found to contain 0.092 g of ^{40}K and 0.530 g of ^{40}Ar. How old is the rock? What assumption(s) are you making to solve this problem? Do you believe they are valid?

13. What are the partial half-lives for the two major modes of decay for ^{36}Cl?

14. A sample of copper known to contain 9.76×10^{-15} g of ^{64}Cu gives a count rate of 2345 cpm on a detector that is 7.20% efficient toward this energy of beta particle. Assuming the detector can only detect beta decay, calculate the half-life of this nuclide.

15. A rock contains 1.77 g of ^{232}Th. What is the total activity, from all radioactive nuclides, of this rock?

16. What is the mass ratio of ^{230}Th to ^{234}Th in a sample of uranium ore?

17. A sample of ^{99}Mo with an activity of 7.67×10^{13} dpm is isolated. After 56 h, what activity of ^{99m}Tc would be observed? Assume that only 86% of all ^{99}Mo decays to ^{99m}Tc, and the remainder decays directly to ^{99}Tc.

18. A 7.0-μg sample of ^{32}Si is isolated. What will the activity of ^{32}P in this sample be after four weeks? Which type of nuclear equilibrium does this represent?

19. How many grams of radium are in a sample of ore containing 1.00 g of Th? What is the activity of ^{219}Rn in this sample?

20. A 7.0-mg sample of ^{48}Cr is isolated. When will its daughter reach maximum activity? Which type of nuclear equilibrium does this represent?

21. A sample was counted 20 times for one minute each in the same geometry. The results (cpm) are tabulated below. Calculate the mean, standard deviation, RSD, and σ_{mean}. How many (%) data are within 1σ, and outside of 2σ? Do these data appear to fit a Gaussian distribution? Do they fit a Poisson distribution? Briefly explain your answers.

2018	2098	2217	2075	2087
2112	2207	2117	2045	2062
2056	2039	2139	2153	2222
2122	2048	2143	2225	2110

3 Energy and the Nucleus

3.1 BINDING ENERGY

Back in Chapter 1 we saw that the strong force holds the nucleus together. The strong force is energy, and energy is required to pull a nucleus apart (thank goodness!). The energy required to separate a nucleus into individual nucleons is the nuclear **binding energy**. Nuclear binding energy could also be defined as the amount of energy released when a nucleus is assembled from its individual nucleons. Energy doesn't just suddenly appear. It has to come from somewhere. In our theoretical construction of a nucleus from protons and neutrons, it comes from mass. You may have learned that mass is always conserved or that energy is always conserved, but the truth of the matter is that mass and energy are collectively conserved. If matter is destroyed, energy is created. Likewise, it is possible to create matter from energy. Sounds like Star Trek, but it is true! The mathematical equation that relates matter and energy is very simple, and very well known:

$$E = mc^2 \qquad (3.1)$$

Energy is equal to mass times the square of the speed of light (3.00×10^8 m/s). When using the equation above, the appropriate unit for mass is the kilogram (kg), which makes the energy unit the joule ($J = kg \cdot m^2/s^2$). Nuclear scientists like to think on the atomic scale, and prefer to use the atomic mass unit (u) as a unit of mass. How does that convert to energy?

$$1 \text{ u} = 931.5 \text{ MeV}$$

One u of mass is equal to 931.5 million electron volts (MeV) of energy. The electron volt (eV) is also a unit of energy, just like the joule. The joule is rather large when thinking on the atomic scale ($1 \text{ J} = 6.242 \times 10^{18}$ eV), so the electron volt is more convenient. An **electron volt** is defined as the amount of energy required to move one electron across a 1-V potential—in other words, not much.

So, what about nuclear binding energy? We can calculate it by looking at how much energy is released when a nucleus is assembled from protons and neutrons. Let's start small, like an alpha particle ($^4_2$He). The mass of a proton is 1.007276 u, and the mass of a neutron is 1.008665 u, so the mass of an alpha particle should be:

$$(2 \times 1.007276 \text{ u}) + (2 \times 1.008665 \text{ u}) = 4.031882 \text{ u}$$

At this point, it'd be nice to compare this mass to the mass of an alpha particle. If we look up ^{4}He in a chart of the nuclides, an *atomic* mass of 4.026032 u is given, which means electrons are included in the mass. Let's add the two electrons to our mass:

$$4.031882 \text{ u} + (2 \times 0.000548 \text{ u}) = 4.032978 \text{ u}$$

The electrons are lightweights, but they do make a difference. This is the expected atomic mass, and it is clearly larger than the actual atomic mass of 4.026032 u, meaning that some mass is converted to energy during the assembly of a ^{4}He atom. This difference in mass is called the **mass defect**. The mass defect and binding energy for ^{4}He are:

$$4.032978 \text{ u} - 4.026032 \text{ u} = 0.006946 \text{ u}$$

$$0.006946 \text{ u} \times \frac{931.5 \text{ MeV}}{\text{u}} = 6.470 \text{ MeV}$$

It should now be apparent why a crazy number of significant figures were necessary at the beginning of these calculations. Once the masses are subtracted, the number of significant figures can go way down (from 7 to 4 in our case!). A good rule of thumb is to express masses six places past the decimal in problems like this. Let's try another one!

Example: Calculate the mass defect and binding energy for ^{56}Fe.

Let's save time and combine the proton and electron mass by using the atomic mass of ^{1}H (1.007825 u).

$$^{56}_{26}\text{Fe}_{30} \quad 26(^1_1\text{H}) \quad 26(1.007825 \text{ u}) = 26.203450 \text{ u}$$

$$30(^1\text{n}) \quad 30(1.008665 \text{ u}) = 30.259962 \text{ u}$$

expected atomic mass	56.463412 u
actual atomic mass	55.934942 u
mass defect	0.528470 u
convert to energy	$\times 931.5$ MeV/u
binding energy	492.3 MeV

Wow! The mass defect and binding energy are enormous for ^{56}Fe when compared to those of ^{4}He. But we really should've expected it. Every nucleon will lose a little mass in forming a nucleus; the more nucleons that are glued together the more total mass will be lost. The more mass lost, the greater the binding energy. This also makes sense from the perspective of taking the nucleus apart. The bigger it is, the more energy should be required to pull it completely apart.

Is there any way to compare the binding energy values for ^{4}He and ^{56}Fe? Yes! Dividing the binding energy (or the mass defect) by the number of nucleons gives the average amount of energy it takes to pluck a nucleon from that nucleus. In other words, how tightly held is each nucleon? This quantity is known as the **binding energy per nucleon**, and can allow comparisons between nuclei.

$$\mathstrut^{4}_{2}\text{He} \quad \frac{6.470 \text{ MeV}}{4 \text{ nucleons}} = 1.618 \frac{\text{MeV}}{\text{nucleon}}$$

$$\mathstrut^{56}_{26}\text{Fe} \quad \frac{492.3 \text{ MeV}}{56 \text{ nucleons}} = 8.791 \frac{\text{MeV}}{\text{nucleon}}$$

It's a heck of a lot harder to pry a nucleon off a ^{56}Fe nucleus than off a ^{4}He nucleus. If these calculations are repeated for all naturally occurring nuclides, a graph like figure 3.1 could be produced. Those nuclides with relatively high values for their binding energy per nucleon could be considered more stable than those with lower values. In fact, if two light nuclides with low values were to combine (fusion) or one heavy nuclide were to split (fission) into two, a great deal of energy could be released as nuclides with greater stability are formed. These two processes will be discussed in more detail in later chapters. For now, it is simply important to understand that changes in binding energy per nucleon are the energetic driving force behind fission and fusion.

The binding energy per nucleon is strongest from $A \approx 50$ to $A \approx 100$, and it is weaker for nuclides above and below that range. Remember that the strong force is about 100 times stronger than the Coulomb repulsion (electromagnetic) of protons. When we add another nucleon to a small nuclide we generally get a lot of strong force with it, and very little Coulomb repulsion, the net result being higher binding energy per nucleon. This works until the nuclide starts getting too big. Then the range of the strong force for each nucleon becomes smaller than the nucleus. Gradually Coulomb repulsion becomes more and more significant, and the nucleons are held less tightly.

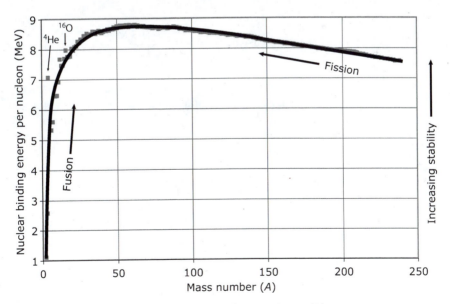

FIGURE 3.1   Relative stabilities of the naturally occurring nuclides.

It is also interesting to look at some of the anomalies in the low mass end of the graph. Notice that ^{4}He sits well above the curve, indicating that its nucleons are bound especially tightly, and therefore has some unusual stability. This is due to the fact that ^{4}He has two protons and two neutrons—both are 'magic numbers' (closed nuclear shell), which imparts extra stability. Other nuclides with closed shells, such as ^{16}O, also exhibit usually high nuclear stability.

Another way to understand the instability of large nuclides toward fission and the instability of small nuclides toward fusion is to look at the changes in mass. When a large nuclide undergoes fission, the products usually have less mass than the reactant(s). Likewise, when some small nuclides combine to form a bigger one, mass is lost. When mass disappears, it is converted to energy.

3.2 TOTAL ENERGY OF DECAY

We'll get to the energetics of fusion and fission in Chapter 8, but the conversion of matter to energy also applies to decay processes. How much energy are we talking · about? How can we calculate it? It is simply a matter of figuring out how much mass is lost during the course of the decay, then converting to energy. To calculate the mass lost, add up the masses of the products and subtract that from the mass of the parent.

Example: How much energy is released when a ^{222}Rn nuclide decays?
First, write out the decay equation:

$$^{222}_{86}\text{Rn} \rightarrow \, ^{218}_{84}\text{Po} + \, ^4_2\text{He}$$

Look up the masses, and do the math:

$$[222.0175570 \text{ u} - (218.008966 \text{ u} + 4.002603 \text{ u})] \times \frac{931.5 \text{ MeV}}{\text{u}} = 5.578 \text{ MeV}$$

For decay processes this energy is the **total energy of decay**. Where does this energy go? Most of it goes into the kinetic energy of the alpha particle. According to the chart of the nuclides, the most common alpha particle emitted by ^{222}Rn has an energy of 5.4895 MeV. Sounds about right. The rest is kinetic energy (recoil) of the daughter. We'll look at this in more detail in Chapter 4.

Keep in mind that 5.578 MeV is the amount of energy released for a *single* decay of ^{222}Rn. What if a whole mole of ^{222}Rn nuclides decayed?

$$\frac{5.578 \text{ MeV}}{\text{decay}} \times \frac{1.6 \times 10^{-13} \text{ J}}{\text{MeV}} \times \frac{6.022 \times 10^{23} \text{ decay}}{\text{mol}} = 5.4 \times 10^{11} \frac{\text{J}}{\text{mol}}$$

For comparison purposes, let's consider a common chemical reaction that cranks out a lot of heat. Burning a mole of methane (CH_4 a.k.a. natural gas) produces only 4×10^8 J, a 1000-fold difference from the decay of ^{222}Rn! Energy is produced by this chemical combustion reaction. Is mass lost? You bet. If you do the math correctly, it should come out to be $\sim10^{-8}$ g. It is such a small amount that it would be difficult to measure. Therefore, mass is essentially conserved in chemical reactions.

Let's take this one step further. If we could harness all of the energy from the decay of one mole of ^{222}Rn as heat energy (it is easy!), how much ice could it melt? We'll have to dredge up some thermochemistry to answer this question. Remember that ice at 0°C still needs heat applied to melt to liquid water at 0°C. This is called the heat of fusion, and is equal to 6.30 kJ/mol for water:

$$5.4\times10^{11}\ \text{J}\times\frac{\text{kJ}}{1000\ \text{J}}\times\frac{\text{mol}}{6.30\ \text{kJ}}\times\frac{18.0\ \text{g}}{\text{mol}}\times\frac{\text{kg}}{1000\ \text{g}}=1.5\times10^{6}\ \text{kg}$$

One and a half million kilograms! That's a lot of ice. Radioactive decay clearly produces a lot of heat. Believe it or not, this helps explain why the earth's core is so hot. Given the age of the earth ($\sim$4.5 billion years), and the rate at which it is cooling (as a rock in space … we're not getting into global warming here!), it should've cooled off a lot more than it has. The answer is that there are and have been lots of radioactive materials inside the earth which have been busy decaying. All the energy from these decays is absorbed by the planet as heat, which helps the planet's core remain toasty warm.

Radioactive decay within the planet also explains where helium comes from. Because of its low atomic mass, helium can escape into space when released into the atmosphere. So where do we get all the helium to fill our balloons? When a radioactive nuclide undergoes alpha decay, the alpha particle quickly picks up a couple of electrons and becomes a helium atom. If this happens below the surface of the earth, the helium can become trapped underground. The helium we fill balloons with comes from wells, usually dug looking for natural gas or oil.

3.3 DECAY DIAGRAMS

Nuclear decay is always exothermic; in other words, it always gives off energy. Nuclear scientists sometimes use the term 'exoergic' to mean the same thing. Nuclides always decay from a less stable nuclide to a more stable one. The atomic number (Z) also changes (except for γ decay), so we can illustrate decay by plotting Z vs. energy, as in figure 3.2.

The horizontal lines in figure 3.2 represent nuclidic states—a nuclide in a particular energy state. The arrow begins at the parent and points toward the daughter. Notice that the daughter is always lower in energy than the parent, indicating that energy is given off during the decay process. The vertical displacement between parent and daughter is the total energy of decay, a.k.a. the energy of transition, and

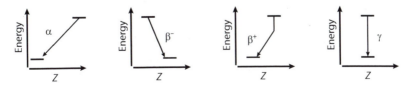

FIGURE 3.2 Generic decay diagrams for alpha, beta, positron, and gamma decay.

corresponds to the mass lost in the decay. The horizontal position of the daughter relative to the parent varies depending on the type of decay. Two protons are lost during alpha decay, so the daughter is two units of Z lower than the parent. A neutron is converted to a proton during beta decay, so the daughter is one unit higher than the parent. The opposite nucleon conversion happens during positron decay, so the daughter ends up one lower in Z. There is no change in atomic number during gamma decay. Drawing these diagrams helps us to see the (sometimes complex) energy changes involved in decay. These diagrams are also known as energy diagrams or decay schemes.

Let's take a look at some examples, starting with the three flavors of beta decay.

Example: ^{35}S decays to ^{35}Cl by emitting a beta particle. Draw a decay diagram, and write out the balanced equation. Also, calculate the total energy of decay.

The decay diagram is drawn in figure 3.3 and represented in the equation below:

$$^{35}_{16}\text{S} \rightarrow ^{35}_{17}\text{Cl} + _{-1}e$$

The atomic mass of ^{35}S is 34.969032 u and the atomic mass of ^{35}Cl is 34.968853 u. The total energy of decay (E) is therefore:

$$(34.969032 \text{ u} - 34.968853 \text{ u}) \times \frac{931.5 \text{ MeV}}{\text{u}} = 0.167 \text{ MeV}$$

According to the chart of the nuclides, the total energy of this decay is 0.167 MeV! Our result is spot on.

Careful examination of figure 3.2 reveals something strange about positron decay. It is the only one with a bent arrow. In every positron decay, the mass of two electrons must be created, which requires 1.022 MeV of energy:

$$[2 \times (5.486 \times 10^{-4} \text{ u})] \times \frac{931.5 \text{ MeV}}{\text{u}} = 1.022 \text{ MeV}$$

Since this is required for every positron decay, we account for it by drawing a line straight down from the parent nuclide. The slanted arrow then represents the energy

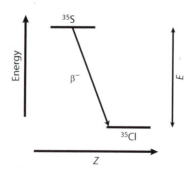

FIGURE 3.3   Decay diagram for ^{35}S.

carried off by the positron. Why is mass created in positron decay, but not in other forms of decay? The answer will have to wait until Chapter 4—stay tuned!

Example: ^{17}F undergoes positron decay to ^{17}O. Write out the balanced equation and draw a decay diagram. If the total energy of decay is 2.7607 MeV and the atomic mass of the daughter is 16.999132 u, what is the atomic mass of ^{17}F?

$$^{17}_{9}F \rightarrow {}^{17}_{8}O + {}_{+1}e$$

The decay diagram is shown in figure 3.4. Notice how the energy value given next to the positron symbol differs from the total energy of decay by 1.02 MeV.

The mass of an atom of ^{17}F can be calculated by taking advantage of the fact that the total energy of decay is related to the difference in mass between parent and daughter, just as it was for regular beta decay, above. Except we're doing this problem backwards, starting at total energy of decay and working back to the masses. The total mass lost in this decay is:

$$2.7607 \text{ MeV} \times \frac{u}{931.5 \text{ MeV}} = 0.002964 \text{ u}$$

which allows us to calculate the mass of the parent:

$$16.999132 \text{ u} + 0.002964 \text{ u} = 17.002096 \text{ u}$$

The actual mass is reported as 17.002095 u. Another excellent result!

Example: ^{41}Ca undergoes electron capture decay to ^{41}K. Write out the balanced equation and draw a decay diagram. If the total energy of decay is 0.4214 MeV and the atomic mass of ^{41}Ca is 40.96228 u, what is the atomic mass of ^{41}K?

The decay diagram is similar to that for positron decay, except that a straight arrow is drawn (fig. 3.5). The balanced equation is below—remember that the electron is an orbital electron:

$$^{41}_{20}Ca + {}_{-1}e \rightarrow {}^{41}_{19}K$$

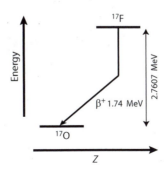

FIGURE 3.4   Decay diagram for ^{17}F.

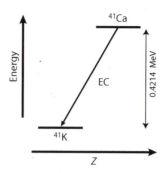

FIGURE 3.5   Decay diagram for ⁴¹Ca.

Calculating the atomic mass of the daughter is very similar to the previous example. Keep in mind that the mass of the daughter is always less than that of the parent.

$$0.4214 \text{ MeV} \times \frac{u}{931.5 \text{ MeV}} = 0.000452 \text{ u}$$

$$40.962278 \text{ u} - 0.000452 \text{ u} = 40.961826 \text{ u}$$

What if we wanted to include more than one decay in our decay diagram? No problem, just draw it in.

Example: Draw the decay diagram for the decay of ³²Si to ³²S. If the total energy of decay for the beta decay of ³²Si to ³²P is 0.224 MeV, and that for ³²P to ³²S is 1.7107 MeV, and the mass of ³²S is 31.972071 u, calculate the atomic mass of ³²Si.

The problem doesn't ask, but the two decays can also be represented as:

$$^{32}_{14}\text{Si} \rightarrow\ ^{32}_{15}\text{P} + _{-1}e$$

$$^{32}_{15}\text{P} \rightarrow\ ^{32}_{16}\text{S} + _{-1}e$$

The decay diagram is shown in figure 3.6. Combining the two decay energies and converting to mass allows us to calculate the atomic mass of ³²Si:

$$31.972071 \text{ u} + \left[(0.224 \text{ MeV} + 1.7107 \text{ MeV}) \times \frac{u}{931.5 \text{ MeV}} \right] = 31.974148 \text{ u}$$

Even though no nucleons change identity and no particle is spit out of the nucleus during gamma decay, the nucleus still manages to lose mass. Energy, in the form of a gamma photon, is produced, and therefore mass must be destroyed in the process.

Example: Write out the decay equation and draw a decay diagram for the decay of ⁷⁷ᵐBr to ⁷⁷Br. If the mass of ⁷⁷ᵐBr is 76.921494 u and ⁷⁷Br is 76.921380 u, calculate the total energy of decay.

Hopefully, this is starting to get a bit easier (fig. 3.7).

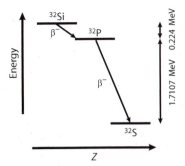

FIGURE 3.6 Decay diagram for the decay of ^{32}Si to ^{32}S.

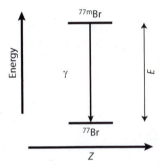

FIGURE 3.7 Decay diagram for ^{77m}Br.

$$^{77m}_{35}\text{Br} \rightarrow {}^{77}_{35}\text{Br} + \gamma$$

$$(76.921494 \text{ u} - 76.921380 \text{ u}) \times \frac{931.5 \text{ MeV}}{\text{u}} = 0.106 \text{ MeV} = 106 \text{ keV}$$

In most of the examples we've seen so far, the total energy of decay is taken away from the nucleus by particles or photons. As noted earlier, some of the energy of alpha decay is left behind with the nucleus as recoil. This introduces a minor complexity to the energetics of alpha decay. It will be discussed in Chapter 4.

QUESTIONS

1. Calculate the binding energy per nucleon for ^{239}Pu. Would you expect this nuclide to undergo, fission, fusion, or neither? Briefly explain.
2. Calculate the average binding energy per nucleon and the average mass of a proton in ^{99}Tc. Would you expect this nuclide to undergo, fission, fusion, or neither? Briefly explain.
3. Your body is 18.6% (by weight) carbon. Using 1.2×10^{-10} as the percent natural abundance of ^{14}C, calculate the activity (dpm) of ^{14}C in your body.

How much energy (J) is being released in your body every minute as a result of ^{14}C decay?

4. Radioactive materials have sometimes been used directly as a power source. Thermal energy, produced during the decay process, is converted to electrical energy using a thermocouple. How much energy (J) could be generated from the decay of a mole of ^{238}Pu? How much energy (J) is generated per minute by 2.0 kg of ^{238}Pu?

5. ^{165}Er undergoes isobaric decay to ^{165}Ho. The atomic mass of the parent is 164.930723 u, and that of the daughter is 164.930319 u. Using only the information given here, write out the decay equation, draw a decay diagram, and calculate the total energy of decay.

6. ^{89m}Y decays to its ground state. The total energy for the decay is 0.9090 MeV, and the atomic mass of the daughter is 88.905848 u. Using only the information given here, write out the decay equation, draw a decay diagram, and calculate the atomic mass of ^{89m}Y.

7. Using the information given in the chart of the nuclides, draw a decay diagram for ^{13}N. Show how its atomic mass can be calculated from the mass of the daughter.

8. Draw a decay diagram for the main decay pathways for ^{68}Ge to a stable nuclide.

4 Radioactive Decay: The Gory Details

WARNING! This chapter is not intended for the casual reader. Proceed only if you are truly curious about better understanding the different forms of decay, or your instructor is telling you that you better read and understand this stuff.

4.1 ALPHA DECAY

This form of decay is most commonly observed for nuclides heavier than ^{208}Pb. The following can generically represent alpha decay:

$$^{A}_{Z}X \rightarrow \, ^{A-4}_{Z-2}Y^{2-} + \, ^{4}_{2}He^{2+}$$

Basically the nucleus is blowing chunks; in this case a ^{4}He nucleus, otherwise known as an alpha particle. The parent is just too big, and the best way it can move toward a stable nuclide is to get rid of four of its nucleons. In terms of stuff spit out by an unstable nucleus (like beta particles, positrons, and photons), alpha particles are quite large. Notice that, in the equation above, charge is also accounted for. Assuming all of the electrons stay on the daughter, it should have a $2-$ charge because two protons just left the nucleus. The $2+$ charge on the alpha particle provides charge balance. It is not always necessary to indicate the electrical charges in decay equations, so long as they are understood.

Alpha decay can also be represented using much simpler notation:

$$^{A}_{Z}X \xrightarrow{\alpha} \, ^{A-4}_{Z-2}Y$$

Because of the relatively large size and high charge of the alpha particle, it has a high probability of interacting with matter. In other words, it is very easy to shield alpha radiation. Alpha particles are also *monoenergetic*, because the parent and daughter have very specific (quantized) energy states. This means that if the energy states of the parent and daughter are always the same, the alpha particle will always have the same energy. The transition from parent to daughter always involves a specific amount of energy, as we've already seen in drawing simple decay diagrams.

The fly in this ointment is the fact that the daughter isn't always produced in the same energy state. If the daughter is in an excited state, then the alpha particle emitted in that decay will have less energy. The excited states are also quantized, so even if several different alpha energies are possible for a decay, only those energies will be observed. How can this be represented in a decay diagram? It'll take a while to answer that. First, let's review a bit of what we learned in Chapter 3.

Example: When ^{246}Cm decays, alpha particles with energies of 5.386 and 5.343 MeV are observed. Gamma photons with an energy of 44.5 keV are also observed. Write out the decay equation and calculate the total energy of decay.

The decay equation (without charges):

$$^{246}_{96}\text{Cm} \rightarrow {}^{242}_{94}\text{Pu} + {}^{4}_{2}\text{He}$$

Calculate the total energy of decay, just like in Chapter 3:

$$[246.067218 \text{ u} - (242.058737 \text{ u} + 4.002603 \text{ u})] \times \frac{931.5 \text{ MeV}}{\text{u}} = 5.475 \text{ MeV}$$

Notice that the total energy of decay is greater than the energies of both alpha particles observed. This should be expected, as part of the total energy of decay goes into the kinetic energy of the daughter. How much? The answer has to do with momentum. Momentum is mass times velocity and is conserved when a projectile is shot out of a larger object. When a bullet is shot from a gun, there is a certain amount of recoil (or kick) in the gun. The gun wants to move in the opposite direction of the bullet. The amount of recoil will depend on the masses of the bullet and gun. The momentum for each will be the same, but the velocity will be higher for the bullet because it has a lower mass. The same is true for alpha decay, as illustrated in figure 4.1.

The alpha particle moves a lot faster because it has a lower mass. Since kinetic energy is 1/2 times the mass times the velocity squared, the alpha particle will also have a greater kinetic energy than the daughter. Classical physics tells us that the kinetic energy of the alpha particle (or any smaller object) can be calculated using the following formula:

$$K_\alpha = K_\text{T}[M/(M+m)] \tag{4.1}$$

where M is the mass of the daughter, m is the mass of the alpha particle, and K_T is the total energy of decay. Likewise, the kinetic energy of the daughter (larger object) can be determined using:

$$K_\text{F} = K_\text{T}[m/(M+m)] \tag{4.2}$$

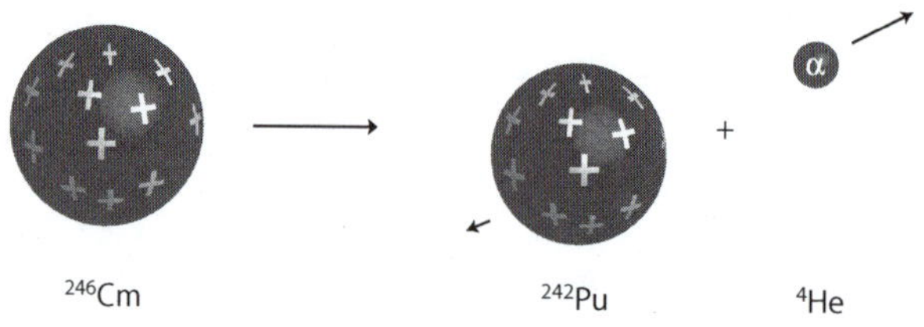

246Cm   242Pu   4He

FIGURE 4.1   Recoil in the alpha decay of ^{246}Cm.

Example: Using the data from the previous example, calculate the maximum energy of the alpha particle and the recoil energy of the daughter in the decay of ^{246}Cm.

$$K_\alpha = 5.475 \text{ MeV}[242/(242+4)] = 5.386 \text{ MeV}$$

$$K_F = 5.475 \text{ MeV}[4/(242+4)] = 0.0890 \text{ MeV} \approx 89 \text{ keV}$$

Using mass numbers for the masses is sufficient for the level of accuracy needed. Check it for yourself—use more exact masses and see if there is a significant effect on the results.

Notice that the two energy values determined above add up to the total energy of decay. Energy is conserved as it is partitioned between the daughter and the alpha particle:

$$K_T = K_\alpha + K_F = 5.386 \text{ MeV} + 0.089 \text{ MeV} = 5.475 \text{ MeV}$$

What does "maximum energy of the alpha particle" mean? It means that the total energy of decay is exactly partitioned between the alpha particle and the daughter. In other words, a ^{246}Cm decay that emits a 5.386 MeV alpha particle corresponds to a transition from the ground state of ^{246}Cm to the ground state of ^{242}Pu. If ^{246}Cm emits an alpha particle with anything less than the maximum energy, then the transition must end at an excited state of ^{242}Pu, not the ground state. We're finally ready to draw the decay diagram for the decay of ^{246}Cm. It is shown in figure 4.2.

The difference in energy between the ground and the excited states of ^{242}Pu must correspond to the difference between the two alpha particle energies:

$$5.386 \text{ MeV} - 5.343 \text{ MeV} = 43 \text{ keV}$$

Keep in mind that these energy values are determined by experiment (empirically determined), and they all have some error associated with them. The difference

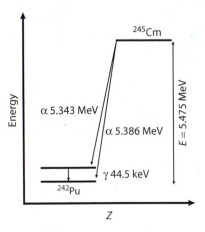

FIGURE 4.2 Decay diagram for ^{246}Cm.

between the calculated 43 kV and observed 44.5 kV is mildly disconcerting, but strongly suggests that the observed gamma photon can occur following the emission of a 5.343 MeV alpha particle from ^{246}Cm.

Finally, notice that the arrows from ^{246}Cm do not quite reach the two nuclear states of ^{242}Pu. This is to show that the total energy of decay is greater than the energy of the 5.386 MeV alpha particle (or the sum of the 5.343 MeV alpha particle and the 44.5 keV gamma photon). Now you know why the alpha arrow didn't touch the daughter in figure 3.2.

Example: Show that mass and energy are collectively conserved in the alpha decay of ^{246}Cm.

Initially, all we have is the mass of the parent. If mass and energy are conserved, then:

$$\text{mass of parent} = \text{mass of daughter} + \text{mass of } \alpha + \text{energy of } \alpha + \text{recoil energy}$$

$$246.067218 \text{ u} = 242.058737 \text{ u} + 4.002603 \text{ u} + (5.386 \text{ MeV} + .089 \text{ MeV})\frac{\text{u}}{931.5 \text{ MeV}}$$

$$= 246.067218 \text{ u}$$

Note that this only works with the maximum energy alpha particle. If we wanted to use the 5.343 MeV alpha, we'd also need to include the 44.5 keV γ-ray on the right-hand side of the equation above.

Let's take a look at how the N/Z ratio changes in the reaction we've been studying.

$$^{246}_{96}\text{Cm}_{150} \xrightarrow{\alpha} {}^{242}_{94}\text{Pu}_{148}$$

$$N/Z = 1.56 \qquad\qquad 1.57$$

It *increases* while forming a lower Z nuclide! As you remember from figure 1.3, it would be better to form a smaller nuclide with a lower N/Z ratio. After a series (or even just one or two) of alpha decays, the ratio will get to the point where beta decay is necessary. Examination of figure 1.5 illustrates this point nicely. The increase in N/Z from alpha decay requires each of the three naturally occurring decay series to be a mix of alpha and beta decay. Notice also that alpha decay does not affect the even or odd nature of the number of protons or neutrons. Both parent and daughter in our example are even-even (*ee*). If the parent had been *eo*, the daughter would also be *eo*.

One final point about alpha decay: just because a nuclide is larger than ^{208}Pb, doesn't mean it'll decay by alpha. Large nuclides with poor N/Z ratios *can* also decay by beta, electron capture, or positron. In some cases branched decay (alpha plus one or more flavors of beta) is observed. Likewise, a nuclide doesn't have to be huge to decay via alpha emission. ^{157}Lu is believed to decay via alpha, and ^{8}Be splits into two alpha particles.

4.2   BETA DECAY

There are some problems with the rather simple way beta decay has been represented in this book. Don't worry, our representation has just been incomplete, not inaccurate. As represented so far, beta decay violates three conservation laws:

1. The Law of Conservation of Particles. The beta particle is created by the nucleus, but from what? Flipping a quark? A particle cannot be created all by itself; an antiparticle must also be created at the same time.

2. The Law of Conservation of Angular Momentum. Angular momentum can be thought of simply as spin. Every subatomic particle and every nucleus has a spin value. It can be a whole number, or a whole number divided by 2. This law states that the difference in spin between reactants and products in a nuclear decay must be a whole number. Let's take a look at the spin values of the known participants in the decay of ^{14}C.

$$^{14}C \rightarrow {}^{14}N + {}_{-1}e$$
$$\text{spin values: } 0 \qquad 1 \qquad \tfrac{1}{2}$$

The sum of the spins on the product side is $1\tfrac{1}{2}$. Since the spin of the parent is 0, the change in spin will also be $1\tfrac{1}{2}$, not a whole number, and therefore is in violation of the law. Written in this way, every beta decay will appear to violate the Law of Conservation of Angular Momentum.

3. The Law of Conservation of Energy. Energy can be converted into different forms of energy or even into matter, but it can't just go away or be created from nothing. The problem with beta decay is that the beta particles are not monoenergetic like in alpha decay. A broad energy distribution is observed for all beta particles, instead of a relatively narrow peak, as illustrated in figure 4.3. It is not possible for the same nuclear transition to take place but different energies be produced.

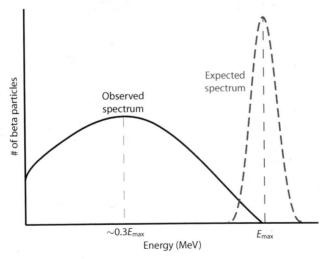

FIGURE 4.3   Expected (dashed line) and observed (solid line) energy spectra for beta particles.

The answer to all of our problems would be an antiparticle that is emitted at the same time as the beta particle. This antiparticle should have a spin value of 1/2 and it should share decay energy with the beta particle in a relatively random fashion. Finally, this antiparticle should have a very low probability of interacting with matter, as it has proven very difficult to detect. An antineutrino ($\overline{\nu}$) has all of these attributes. It has no charge, very low mass, and an extremely low probability of interacting with matter. We can now write a complete equation for the decay of ^{14}C:

$$^{12}_{6}C \rightarrow {}^{14}_{7}N + {}_{-1}e + \overline{\nu}$$

We can also write a generic beta decay reaction (including charges) as:

$$^{A}_{Z}X \rightarrow {}^{A}_{Z+1}Y^{1+} + {}_{-1}e^{-1} + \overline{\nu}$$

The plus one charge on the daughter (Y) results from it having the same number of electrons as the parent (X), yet one more proton. The negative one charge on the beta particle balances this out. In general, beta decay reactions will not be written in this text with explicit charges or antineutrinos. The reader should understand they are implied. The main reason for leaving the antineutrino off is that it remains a rather elusive and poorly understood antiparticle.

Beta decay reactions can also be written in shorthand as:

$$^{A}_{Z}X \xrightarrow{\beta^-} {}^{A}_{Z+1}Y$$

The ultimate example of beta decay is the decay of free neutrons. On their own, neutrons have a half-life of $\sim$10 min. It is a good thing they are stable when incorporated into nuclei!

$$^{1}_{0}n \rightarrow {}^{1}_{1}p + {}_{-1}\beta^- + \overline{\nu}$$

All of the stuff in this reaction has a spin value of 1/2. Is the Law of Angular Momentum violated in this decay?

The maximum beta energy (E_{max}) is given on the chart of the nuclides, and is often quoted as the energy of the beta particle. Theoretically, this could only happen if the energy of the antineutrino emitted in the same decay had no energy. Examination of figure 4.3 shows that beta particles never have this energy, since it always shares some with the antineutrino. According to figure 4.3, the most probable energy for a beta particle is $\sim 0.3 E_{max}$.

We've already seen how to draw an alpha decay diagram that incorporates a gamma emission. The same can be done with beta decay. In fact, it is very unusual to observe alpha or beta decay without gamma emission.

Example: ^{47}Sc decays with the emission of two beta particles with E_{max} values of 0.439 MeV and 0.600 MeV. A γ-ray with an energy of 159.4 keV is also observed. Draw a decay diagram that accounts for all of this information.

The complete decay equation is:

$$^{47}_{21}\text{Sc} \rightarrow {}^{47}_{22}\text{Ti}^+ + {}_{-1}e + \overline{\nu}$$

The total energy of decay is calculated in the usual manner:

$$(46.952408 \text{ u} - 46.951764 \text{ u}) \times \frac{931.5 \text{ MeV}}{\text{u}} = 0.600 \text{ MeV}$$

Since we are using the atomic masses of the parent and daughter nuclides in the calculation above, we don't need to add in the mass of the beta particle—it is already included in the atomic mass of ^{47}Ti. The mass of the antineutrino is negligible, so it does not need to be included in the calculation.

The difference in energy between the two beta particles is close to the energy of the observed γ-ray:

$$0.600 \text{ MeV} - 0.439 \text{ MeV} = 0.161 \text{ MeV} = 161 \text{ keV}$$

We're ready to draw! The 0.600 MeV beta particle is equal in energy to the total energy of decay, and therefore corresponds to a transition between the ground states of the parent and daughter. The sum of the lower-energy beta particle and the energy of the γ-ray is pretty close to the total energy of decay, so odds are good that the 0.439 MeV beta emission leads to an excited state of ^{47}Ti, which is 0.1594 MeV higher in energy than its ground state. The complete diagram is shown in figure 4.4.

The most common decay pathway(s) for a radioactive nuclide does (do) not always involve a direct transition between ground states, and can involve more than one gamma emission. The most prominent beta particle observed in the decay of ^{43}K has an energy of 0.83 MeV, but its total energy of decay is 1.82 MeV. As part of this pathway, gamma photons with energies of 372.8 and 617.5 keV

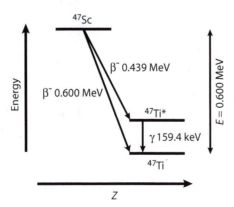

FIGURE 4.4   Decay diagram for ^{47}Sc.

are also observed. The three emitted energies add up to the total energy of decay:

$$0.83 \text{ MeV} + 0.3728 \text{ MeV} + 0.6175 \text{ MeV} = 1.82 \text{ MeV}$$

The primary decay pathway for ^{43}K must therefore look like figure 4.5. Based on the information provided, we can't decide which order the two gamma emissions should be in. The only way to know for sure is to consult a higher authority. A convenient source is the chart on-line at http://atom.kaeri.re.kr/ton/. They conveniently provide a decay diagram, which shows that the decay of ^{43}K involves a few other minor pathways as well as affirming figure 4.5 as correct for the major pathway.

A significant part of the discussion of alpha decay earlier in the chapter was the issue of recoil. Thus far into our look at beta decay, it hasn't been mentioned. Are the calculations above done properly? Let's find out.

Example: Calculate the recoil energy resulting from beta decay for a free neutron.

This is a good example because a neutron is relatively lightweight and the total energy of decay is equal to E_{max} (0.782 MeV). If recoil is to be observed in beta decay, we should see it here. We can use the same equations as for alpha decay, except m is now the mass of an electron:

$$K_{\text{F}} = K_{\text{T}} \left(\frac{m}{M+m} \right) = 0.782 \text{ MeV} \times \frac{0.000548 \text{ u}}{1.01 \text{ u}} = 0.000424 \text{ MeV}$$

The recoil energy in beta decay is very small compared to the total energy of decay, and can easily be neglected in beta decay energy calculations. The reason beta decay is so different from alpha decay is because the mass of an electron is so much smaller than the mass of an alpha particle.

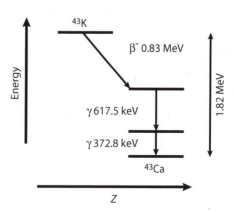

FIGURE 4.5   Partial decay diagram for ^{43}K.

4.3 POSITRON DECAY

Just like beta (minus) decay, positron decay has been simplified in earlier chapters of this book. Without the addition of a neutrino to the product side of the decay reaction, positron decay would also violate the same conservation laws as beta decay. The complete, generic representation of positron decay is:

$$\,^A_Z X \rightarrow \,^A_{Z-1} Y^- + \,_{+1}\beta^+ + \nu$$

A neutrino is formed here, rather than the antineutrino formed in beta decay. This is because the positron itself is an antiparticle, therefore a particle also needs to be created to properly conserve particles. The positron is actually the corresponding antiparticle to the electron. Every particle (like protons, neutrons, neutrinos) has a corresponding antiparticle. These antiparticles are also known as antimatter. That's the fuel for starships in the stunningly successful *Star Trek* sagas. It is also used to provide the devastatingly explosive force in their photon torpedoes. As is often the case, there is some basis in fact here. When antimatter comes into contact with (regular) matter, they annihilate each other, forming energy. When an electron and a positron annihilate each other, two *annihilation photons* are formed, each with an energy of 0.511 MeV. Let's do the math.

Example: Calculate the energy released from a positron/electron annihilation. The reaction is:

$$\,_{+1}e + \,_{-1}e \rightarrow 2\gamma$$

The electron and the positron both have a mass of 5.486×10^{-4} u, and the photons have no mass. Therefore $2(5.486 \times 10^{-4}$ u) is destroyed, and the energy produced is:

$$2(5.486 \times 10^{-4})\,\text{u} \times \frac{931.5\ \text{MeV}}{\text{u}} = 1.022\ \text{MeV}$$

This energy is divided equally between the two annihilation photons, meaning each will be 0.511 MeV. This is a lot of energy released by a small amount of mass. It wouldn't take much antimatter to make a pretty powerful explosive. Interestingly, the two annihilation photons are emitted exactly 180° apart. This forms the basis of positron emission tomography (PET), which is discussed in more detail later in the text. Simultaneous detection of the two photons from annihilation allow for stunning 3-D images of people's innards for medical diagnostic procedures.

Let's get back to the generic positron decay at the top of this page. Paying close attention to charge, the daughter (Y) is formed as a monoanion (−1 charge), assuming it has the same number of electrons as the parent. This negative charge is balanced by the positive charge of the positron, so charge is balanced in this reaction.

What about mass balance? Since the daughter is produced as an anion *and a* positron is produced, the total mass on the product side is the atomic mass of the daughter plus the mass of two electrons. Compare this with beta decay. There, the difference in mass is simply the difference in atomic masses of the parent and daughter nuclides.

This means we have some additional energy accounting to do in positron decay. The total energy of decay is equal to the (kinetic) energy of the positron plus the energy to 'create' two electrons (1.022 MeV—same as you get when they are destroyed!). In beta decay the total energy of decay can equal the maximum possible beta energy for the transition, something that can never be true in positron decay. In order to calculate maximum possible positron energy for a particular decay, we have to account for the extra mass:

$$[\text{at. mass X} - (\text{at. mass Y} + \text{mass}_{-1}e + \text{mass}_{+1}e)] \times \frac{931.5 \text{ MeV}}{\text{u}} \qquad (4.3)$$

Example: ^{18}Ne decays with the emission of two positron particles with E_{max} values of 3.42 and 2.38 MeV. A γ-ray with an energy of 1.04 MeV is also observed. Draw a decay diagram that accounts for all of this information. Also calculate the maximum possible positron energy for this decay.

The total energy of decay can still be calculated in the usual manner:

$$(18.005697 \text{ u} - 18.000938 \text{ u}) \times \frac{931.5 \text{ MeV}}{\text{u}} = 4.433 \text{ MeV}$$

The maximum energy for a positron is:

$$[18.005697 \text{ u} - (18.000938 \text{ u} + 5.486 \times 10^{-4} \text{ u} + 5.486 \times 10^{-4} \text{ u})] \times \frac{931.5 \text{ MeV}}{\text{u}}$$

$$= 3.411 \text{ MeV}$$

Notice it matches one of the E_{max} values given in the problem quite nicely! This means the 3.42 MeV positron gives ^{18}F in its ground state. We can also guess that the 1.04 MeV gamma photon is emitted following the 2.38 MeV one because:

$$3.42 \text{ MeV} - 2.38 \text{ MeV} = 1.04 \text{ MeV}$$

The maximum energy positron plus the energy to 'create' two electrons should equal the total energy of decay:

$$3.42 \text{ MeV} + 1.022 \text{ MeV} = 4.44 \text{ MeV}$$

Close enough. The decay diagram is given in figure 4.6. The short vertical line from ^{18}Ne represents the 1.022 MeV required by all positron decays. If the total energy of decay is less than 1.022 MeV, positron decay is not possible.

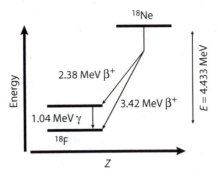

FIGURE 4.6   Decay diagram for ^{18}Ne.

What about recoil? The results should be the same as with regular beta, so the answer is no, it is insignificant. Also like regular beta decay, positron decay can be represented by the following shorthand:

$$_{Z}^{A}X \xrightarrow{\ \beta^+\ } \ _{Z-1}^{A}Y$$

4.4   ELECTRON CAPTURE

Electron capture probably seems like the strangest form of radioactive decay ever, as it is the only one that is not strictly a nuclear change since it requires an orbital electron. Its generic representation is quite similar to that of positron decay:

$$_{Z}^{A}X + \ _{-1}^{0}e \rightarrow \ _{Z-1}^{A}Y^- + \nu$$

Remember that the electron on the left side of the arrow is an orbital electron. The neutrino on the product side balances the loss of a particle on the reactant side. Because the electron is not originally part of the nucleus, some prefer to write electron capture reactions with the following shorthand:

$$_{Z}^{A}X \xrightarrow{\ EC\ } \ _{Z-1}^{A}Y + \nu$$

Example: Draw a decay diagram for ^{72}Se. It decays via electron capture with a total energy of decay of 0.33 MeV. A γ-ray with an energy of 46 keV is observed in all decays.

Since the gamma photon is observed in all decays, the electron capture must lead to an excited state of the daughter, which then decays to the ground state by spitting out the gamma (fig. 4.7).

It can sometimes be difficult to decide how to draw the electron capture arrow. In the example above it is unambiguous, but if information from the printed chart were used, you wouldn't know that the gamma photon is observed in every decay.

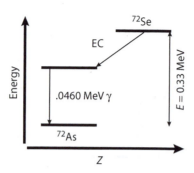

FIGURE 4.7   Decay diagram for ^{72}Se.

In that case, it is best to seek additional information from the online version of the chart (http://atom.kaeri.re.kr/ton/).

Positron and electron capture decays both do the same thing. They both produce a daughter with one less proton and one more neutron than the parent. Can we tell if a particular nuclide will prefer to decay by electron capture or positron emission? Sort of. Electron capture is more likely with high-Z nuclides, while positron decay is more likely for light nuclides, but that's the best we can do. For most proton-rich nuclides, either or both forms of decay are observed.

To better understand electron capture, we'll need to know a bit about orbital electrons. Electrons occupy a volume of space (orbital) around the nucleus that is (more or less) defined by the solution to the mathematically complex Schrödinger equation. The solution generates a set of four numbers that are unique for each electron in an atom. These numbers are called quantum numbers, and individually represent main energy level, shape, orientation, and spin. We'll only be concerned with the quantum number that describes the main energy level. This quantum number is known as the principal quantum number, and is represented by n. It has positive integral values (1, 2, 3, …). Since we're not concerned with orbital shape or orientation, we can represent the different energy levels (shells) as concentric circles around the nucleus. This type of atomic model is called the Bohr model, named after the Danish physicist Niels Bohr who first proposed it. An example is pictured in figure 4.8. The energy levels $n = 1$, 2, and 3, are often represented by the letters K, L, and M, respectively. The letter designations are older, but are more commonly used in nuclear medicine, so we will continue their use in this text. The L shell is further subdivided into three subshells (L_1, L_2, L_3), and M into five. There is only one subshell for the K shell, so it is simply represented as K.

Which electrons crash into the nucleus? Electron capture is most likely to take place with a K shell (K capture) electron. This makes sense, as the K shell is the lowest energy shell, and these electrons spend most of their time fairly close to the nucleus. Of all the electron shells they spend the most time closest to the nucleus. Electron capture from the L (L capture) or M (M capture) shells is also possible, but not probable. For example, K capture happens in ~90%, L capture in ~9%, and M capture in ~1% of all electron capture decays of ^{41}Ca. L and M capture increase in probability with Z, which can be rationalized by the fact that all electron shells

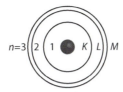

FIGURE 4.8 The Bohr atomic model.

move closer to the nucleus with increasing Z. The same rationale can be applied to the relative probabilities of β^+ and electron capture. A larger Z means the K shell electrons are closer to the nucleus, so electron capture tends to be more popular with heavier nuclides. In reality, it is a bit more complex. For instance, a large value for the total energy of decay tends to favor positron decay, and high decay energies are often observed for low-Z nuclides.

Superficially, electron capture could be considered a stealth mode of decay. If electron capture occurs without a gamma photon, how can we tell decay has taken place? According to the generic decay equation at the beginning of this section, only a neutrino is emitted, and neutrinos are extremely difficult to detect. Fortunately, the atom's electrons can tell us. After K capture has occurred, the K shell will be short one electron. This electron vacancy means that the daughter is created with its electrons in an excited state. To get back to the ground state, an electron from a higher level will drop down to fill the vacancy. When this happens, energy must be emitted. Because the electron energy levels are quantized, the energy emitted for a particular atom is always exactly the same, so long as the transition is between the same two levels. Because the K shell is so much lower in energy than the others, the energy given off when a K vacancy is filled is an X-ray photon.

Like γ-rays, X-rays are high-energy photons (see fig. 1.1). γ- and X-ray photons can have the same energy, and to a detector they are indistinguishable. The difference between the two lies in how they are generated. γ-rays are generated in the nucleus, as a result of a transition from an excited nuclear state to one lower in energy. X-rays are generated by electrons in one of two ways. (1) As we've seen here, an outer shell electron fills an inner shell vacancy (there are other ways to create these vacancies). This type of X-ray is termed a **characteristic X-ray**, because they are produced with specific energies. (2) High-speed (free) electrons can interact with matter by slowing down as it travels near a nucleus. This type of X-ray is called **continuous X-ray**, because the amount of energy released will depend on how close the electron gets to the nucleus, producing a broad spectrum of photon energies. It is also known as **bremsstrahlung**, which is German for 'slowing-down radiation'.

We can finally draw a picture of what happens after electron capture (fig. 4.9). A K shell vacancy is created by the electron capture decay. An L shell electron then fills this vacancy. Keep in mind that any higher energy shell could fill the vacancy, however an L shell electron is the most probable. As a result of the electronic transition, a characteristic X-ray is emitted that has an energy corresponding to the difference in electron-binding energy between the L and K shells in the daughter.

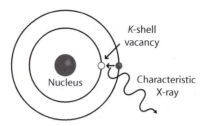

FIGURE 4.9   The aftermath of electron capture—Part I.

Characteristic X-rays are designated according to the subshells that mark the beginning and end of the electronic transition. The most commonly observed transition is labeled $K_{\alpha 1}$, which corresponds to an electron transition from the L_3 subshell to the K shell. $K_{\alpha 2}$ is the $L_2 \rightarrow K$ electronic transition, and $K_{\alpha 3}$ is the $L_1 \rightarrow K$ transition. The $K'_{\beta 1}$ represents a combination of all the transitions from the M shell to the K shell. There is generally little difference in the energies of the different M subshells, so there is also very little difference in the individual K_β values. The energy values for characteristic X-rays correspond to differences in energy between the (sub)shells. These energy values are called **electron-binding energies** because they represent the amount of energy required to remove an electron from an atom. These energy levels can be represented in an energy level diagram such as figure 4.10. This diagram is for the element holmium. Similar diagrams for other elements will vary—typically the numbers will increase with atomic number. This makes sense; it should be more difficult to remove a K shell electron from an element with 92 protons than one with only eight protons. An approximate energy value is given for the five M subshells in figure 4.10. They actually range in value from 1.35 to 2.13 keV.

Example: Using figure 4.10, calculate the energy values of the following characteristic X-rays for holmium: $K_{\alpha 1}$, $K_{\alpha 2}$, and $K'_{\beta 1}$.

$$
\begin{array}{llll}
K_{\alpha 1} & L_3 \rightarrow K & 55.618\ \text{keV} - 8.072\ \text{keV} = 47.546\ \text{keV} \\
K_{\alpha 2} & L_2 \rightarrow K & 55.618\ \text{keV} - 8.918\ \text{keV} = 46.700\ \text{keV} \\
K'_{\beta 1} & M \rightarrow K & 55.618\ \text{keV} - 1.71\ \text{keV} = 53.91\ \text{keV}
\end{array}
$$

Instead of emitting a characteristic X-ray, an atom can get rid of the energy generated by the electronic transition by spitting out one of its electrons, as illustrated in figure 4.11. This process is known as the **Auger** (oh-ZHAY) **effect**, and the electron that gets booted is known as an **Auger electron**.

Auger electrons are designated by: (1) the shell or subshell where the vacancy was originally created; (2) the subshell of the electron that drops to fill the vacancy; and (3) the subshell of origin for the Auger electron. For example, if the electron dropping into the K shell in the left side of figure 4.11 is from the L_2 subshell, and the Auger electron is from the L_3 subshell, then the emitted electron would be called a KL_2L_3 Auger electron. The kinetic energy of the KL_2L_3 Auger electron is equal to the energy gained by the atom when the electron drops from the L_2 subshell to the K shell, minus the energy required to remove an L_3 electron. For holmium it is:

$$KL_2L_3 = 55.618\ \text{keV} - 8.918\ \text{keV} - 8.072\ \text{keV} = 38.628\ \text{keV}$$

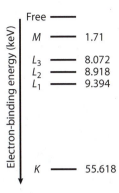

FIGURE 4.10   Energy level diagram for Ho.

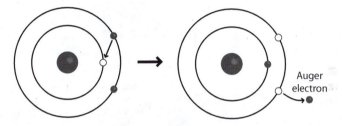

FIGURE 4.11   The aftermath of electron capture—Part II: the Auger effect.

The characteristic X-ray and Auger electron calculations done above are reversed from how they would normally be performed. Typically the energies of the characteristic X-rays and Auger electrons are determined experimentally, and the electron-binding energies are calculated.

Multiple Auger effects are possible in a single decay, potentially leaving the daughter a highly charged cation. Since chemical bonds often involve the sharing of electrons between two atoms, electron capture can be especially disruptive when it occurs within a molecule.

Is it possible to tell whether characteristic X-rays or Auger electrons will be emitted as a result of the electron capture decay? Not really. Both are always observed when electron capture is a decay mode for a particular nuclide. The probability that X-rays will be produced following electron capture is called **fluorescent yield**. It is symbolized by the Greek letter omega (ω). With a K subscript (ω_K) it refers to fluorescent yield when the original vacancy is in the K shell. Figure 4.12 shows a graph of ω vs. atomic number (Z). The probability that characteristic X-rays will be emitted increases with atomic number. In other words, the probability that Auger electrons will be emitted decreases with atomic number. This makes sense; more protons in the nucleus means that electrons in the same shell will be bound more tightly.

4.5   MULTIPLE DECAY MODES

We've already seen that proton-rich nuclides can decay by both positron emission and electron capture. Some nuclides, especially those with moderate atomic numbers, in

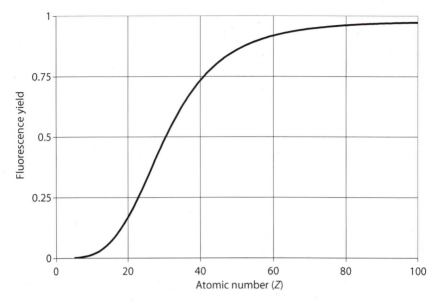

FIGURE 4.12   Fluorescent yield as a function of atomic number. (Data from Hubbell, J.H., *J. Phys. Chem. Ref. Data*, 23, 339, 1994.)

the middle of the belt of stability, and odd numbers of both protons and neutrons, can decay by all three beta modes. Can a decay diagram be drawn for these schizophrenic nuclides? You bet.

Example: ^{80}Br decays via all three beta modes. Using the information provided here, draw its decay diagram. The total energy of decay for β^- is 2.00 MeV, and 1.87 MeV for β^+ and electron capture. The most common beta particles emitted have E_{max} values of 2.00 and 1.38 MeV, while the most common positron has an E_{max} of 0.85 MeV. A 616 keV gamma photon is also prominently observed during β^- decay.

The decay diagram is shown in figure 4.13. All energy values are given in million electron volts (MeV). This diagram can be confirmed from the online chart of the nuclides (http://atom.kaeri.re.kr/ton/). The online chart also provides branch ratios of 91.70% for β^- decay, 6.07% for electron capture decay, and 2.20% for positron decay.

Multiple decay modes are also often observed for heavy ($Z > 82$) nuclides with poor N/Z ratios. These nuclides typically decay via alpha and some form(s) of beta. Note that heavy, neutron-rich nuclides are more likely to decay via β^- than heavy, proton-rich nuclides decaying via electron capture or β^+ emission. This is likely due to the fact that alpha decay also increases the N/Z ratio (see Section 4.1).

4.6   THE VALLEY OF BETA STABILITY

All three flavors of beta decay involve no change in the mass number—they occur along isobars. Let's take a closer look at a set of isobars, in particular the eight nuclides closest to the belt of stability when $A = 99$ (table 4.1).

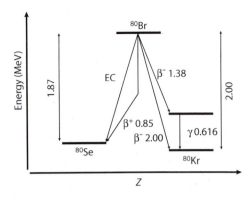

FIGURE 4.13 Decay diagram for ^{80}Br.

TABLE 4.1
Isobars with $A = 99$

Isobar	N	Z	N/Z	Decay Mode(s)	Decay Energy (MeV)
$_{47}$Ag	52	47	1.11	β^+, EC	5.4
$_{46}$Pd	53	46	1.15	β^+, EC	3.39
$_{45}$Rh	54	45	1.20	EC, β^+	2.04
$_{44}$Ru	55	44	1.25	Stable	Stable
$_{43}$Tc	56	43	1.30	β^-	0.294
$_{42}$Mo	57	42	1.36	β^-	1.357
$_{49}$Nb	58	41	1.41	β^-	3.64
$_{40}$Zr	59	40	1.48	β^-	4.56

The neutron number, atomic number, and N/Z ratio are tabulated for each isobar. The optimal value for the N/Z ratio is likely to be close to 1.25 for $A = 99$, since ^{99}Ru is the only stable isobar. The decay mode(s) are given in the next column in table 4.1. Notice that the isobars above ^{99}Ru have lower N/Z ratios (are proton-rich), and all undergo positron and electron capture decay. The more commonly observed decay mode of the two is listed first for each nuclide. The apparently even mix of the two is to be expected for the relatively moderate Z values for these nuclides. The isobars listed below ^{99}Ru all have higher N/Z ratios and decay via β^- emission.

Finally, the total energy of decay is listed for each isobar in table 4.1. Notice that the value for ^{99}Tc is rather small, especially when compared to ^{99}Rh. This suggests that the optimal N/Z ratio for $A = 99$ is probably between 1.25 and 1.30 (the values for ^{99}Ru and ^{99}Tc, respectively), although it is likely closer to 1.25. Notice that the energy values increase the further from ^{99}Ru the isobar gets. We'll soon see that this is not true for all mass number values, but it is often true when A is odd. We can illustrate all of the isobaric decays listed in table 4.1 in a simplified decay diagram, such as figure 4.14.

The decays represented in figure 4.14 are simplified such that only the total energy of decay is represented. In other words, if excited daughter states are formed

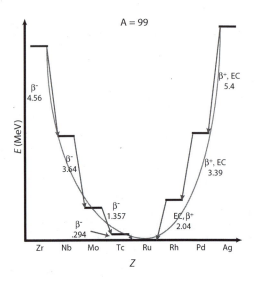

FIGURE 4.14   A simplified decay diagram for some of the $A = 99$ isobars.

in some of these decays, we aren't drawing them; we're only looking at the ground states for each isobar. Since ^{99}Ru is the only stable isobar for $A = 99$, any radioactive isobars formed will eventually decay to ^{99}Ru. Because of the increasing magnitude of the decay energies as the isobars move away from ^{99}Ru, the energy changes for the whole set of isobars can be approximated with a parabola. This mathematical model is known as the **Semi-Empirical Mass Equation** and was (remarkably!) developed by C. F. von Weizsäcker in 1935. He used the word "Mass" because of the simple direct relationship between mass and energy ($E = mc^2$). We could've just as easily graphed mass on the y-axis of figure 4.14, and it would look just the same.

Energy (or mass) is really the third dimension to the chart of the nuclides (the other two are N and Z). If we look at the chart as a whole in this way, we'll see that the stable nuclides are all located at the bottom of an energy valley. The belt of stability is really a valley of stability, and figure 4.14 is a cross-sectional slice of one part of this valley. Figure 4.15 looks at this valley for all known nuclides with mass numbers of 46–71.

The walls of this valley become increasingly steep as Z decreases. This means that the energy changes become more dramatic for lower Z isobars. This is often cited as a reason why proton-rich, low-Z nuclides decay via positron emission rather than electron capture. Apparently, positron decay tends to be favored when the energy differences are great. This can be observed (anecdotally) in table 4.1. ^{99}Rh has the lowest total energy of decay of all the proton-rich nuclides (high N/Z), and it is the only one to decay via electron capture more often than by positron emission. Don't count on this always being true.

As we scan down through the isobars in table 4.1, they alternate between even–odd (*eo*) and odd–even nuclides (*oe*). With either N or Z odd (mass number is odd), it is not surprising that there is only one stable nuclide for $A = 99$. It turns out this is always true. For all sets of known isobars with an odd mass number, there is only

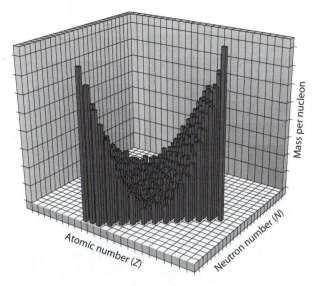

FIGURE 4.15   A 3-D chart of the nuclides from $A = 46$ to $A = 71$.

TABLE 4.2
Isobars with $A = 104$

Isobar	N	Z	N/Z	Decay Mode(s)	Decay Energy (MeV)
$_{49}$In	55	49	1.12	β^+	7.9
$_{48}$Cd	56	48	1.17	EC, β^+	1.14
$_{47}$Ag	57	47	1.21	β^+, EC	4.279
$_{46}$Pd	58	46	1.26	Stable	Stable
$_{45}$Rh	59	45	1.31	β^-, EC	2.44, 1.141
$_{44}$Ru	60	44	1.36	Stable	Stable
$_{43}$Tc	61	43	1.42	β^-	5.60
$_{42}$Mo	62	42	1.48	β^-	2.16
$_{49}$Nb	63	41	1.54	β^-	8.1

one stable nuclide per set of isobars. Check it out for yourself; it is uncanny! Ok, maybe it's not so strange; perhaps one stable nuclide for each value of A is intuitively expected. It shouldn't be. When A is even, there can be 1, 2, or 3 stable nuclides per set of isobars. Check out table 4.2 for $A = 104$.

When the mass number is even, then the nuclides must be odd–odd (*oo*) or even–even (*ee*). With both N and Z odd, it is unlikely the nuclide will be stable. In fact it'll definitely be unstable for isobars with $A = 104$. On the other hand, an *ee* nuclide would be expected to exhibit greater stability relative to nearby nuclides with odd numbers of N and/or Z. Scan down through table 4.2 and the N and Z values. Notice that they alternate between *oo* and *ee*. It's as if there are really two sets of isobars in table 4.2: those with extra (relative!) stability (*ee*) and those with extra instability(*oo*).

The dualistic nature of even isobars is even more apparent when a simplified decay diagram is drawn (fig. 4.16). There are clearly two parabolas that converge at higher energy values. Even isobars are also approximately modeled by von Weizsäcker's equations. Notice also that the two stable nuclides in this set of isobars do not have exactly the same energy. Even among stable nuclides, stability is relative. So what is the optimal N/Z value for $A = 104$? It is between the values for the two stable nuclides: 1.26 (^{104}Pd) and 1.36 (^{104}Ru). It is likely closer to 1.36, as ^{104}Ru is lower in energy than ^{104}Pd. As expected, it is higher than the optimal N/Z value for $A = 99$.

As mentioned earlier, even sets of isobars can include 1, 2, or 3 stable nuclides. Each are presented in figure 4.17, using simplified decay diagrams to illustrate all the observed pathways. There are two ways a set of even isobars can have just one stable nuclide; one with a stable oo nuclide (fig. 4.17a) and the other with a stable ee nuclide (fig. 4.17b). A stable oo nuclide should set off an alarm in your head. Remember there are (only) four stable oo nuclides ($^{2}_{1}$H$_{1}$, $^{6}_{3}$Li$_{3}$, $^{10}_{5}$B$_{5}$, and $^{14}_{7}$N$_{7}$). In every case, they are low-Z nuclides, and are the only known stable nuclide for their set of isobars. The anomalous nature of their stability can be rationalized as due to the steepness of the valley of stability for low-Z nuclides. These nuclides clearly have the best N/Z ratios for their set of isobars, and the adjacent ee nuclides are simply too high in energy to be stable.

$A = 104$ is an example of the most popular isobaric decay pathways when A is even; 83 sets of isobars are known with two stable ee nuclides (fig. 4.17c). A close second are 78 sets with one stable ee nuclide (fig. 4.17b), and a distant last are three sets with three stable ee nuclides (fig. 4.17d). Odds are pretty good that when A is even, there'll be either one or two stable ee nuclides. Close examination of the two and three stable isobar cases illustrated in figure 4.17c and d reveals that the stable ee nuclides are always separated by one unstable oo nuclide. Looking over the last couple of pages suggests that two stable isobars will never be adjacent to each other.

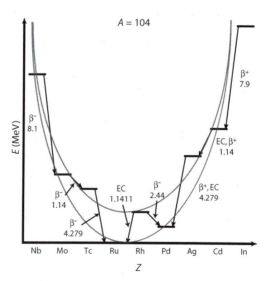

FIGURE 4.16   A simplified decay diagram for some of the $A = 104$ isobars.

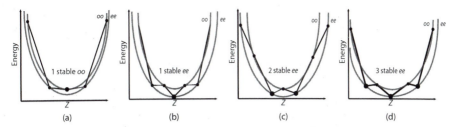

FIGURE 4.17 Simplified decay diagrams representing all possible pathways when mass number (A) is even.

This is known as **Mattauch's Rule**. It is often stated as "stable, neighboring isobars do not exist." Take a look at the chart of the nuclides and see if you can find any exceptions.

4.7 ISOMERIC TRANSITIONS

Many nuclear science textbooks define isomers as only those excited nuclear states with measurable lifetimes. This is a rather arbitrary, and somewhat inconsistent, definition. "Measurable" has changed over the years, and therefore what was once considered only an excited state is now considered an isomer. It likely stems from the pre-Internet era when the chart of the nuclides was only available in printed form. It is not possible to show every excited state for every nuclide in a printed format. The 16th (2002) edition of the chart provides information only on excited states with half-lives greater than one second. To avoid confusion, the chart refers to these as "metastable" states, or simply as "metastates." Recall that this is indicated in the nuclide symbol by adding an m at the end of the mass number, for example, ^{99m}Tc is a metastate of ^{99}Tc.

We'll use a more inclusive definition for isomer. As stated in Chapter 1, isomers are two nuclides with the same number of protons and neutrons, but are in different energy states. An **isomeric transition** is therefore one isomer changing into another through a nuclear change.

Gamma emission is one example of an isomeric transition. Since the energy levels of the isomeric states are quantized, the energy of the gamma photon emitted is exactly equal to the difference in energy of the two nuclear states. Like alpha particles, gamma photons are monoenergetic. Unfortunately, many isomeric states are often possible, which can lead to quite a few different gamma photons. Also recall that γ-rays have no charge or mass, and are strongly penetrating due to their low probability of interacting with matter.

^{125m}Xe decays as pictured in figure 4.18. First it transitions to another excited state of ^{125}Xe, which is 140 keV in energy below ^{125m}Xe and 112 keV above ground-state ^{125}Xe. Curiously, a γ-ray with an energy of 252 keV (140 keV + 112 keV) is not observed. Why not? It is not "allowed" because of the rather dramatic nuclear changes that would have to accompany such a transition. It has to do with the fractions and the + and − signs next to each isomeric state in figure 4.18.

The numbers give the nuclear spin of each state. Each nucleon has a spin of 1/2, and when they are combined together in a nucleus, the nucleus will also have a spin. When

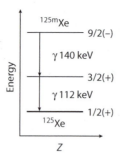

FIGURE 4.18   Isomeric transition of ^{125m}Xe to ^{125}Xe.

A is odd, the ground-state nuclear spin is always $x/2$, where $x = 1, 2, 3, \ldots$ When A is even, the ground-state nuclear spin is 0 or a positive integral value. Nucleons have a tendency to pair up with others of their own kind in the nucleus. When this happens, their spins cancel each other out. As a result, the ground states of most nuclides have spin values that are fairly low. In fact all *ee* nuclides have a 0 spin in the ground state.

As is apparent in figure 4.18, different energy states of the same nuclide have different spin states. Differences in spin are the main difference between isomers. In order to transition from one isomer to another, the nucleus needs to change its spin. Generally speaking, the greater the change in spin state, the less likely the transition will take place. In order for ^{125m}Xe to decay directly to the ground state, the spin state must change from 9/2 to 1/2. Unlikely. Instead it changes to 3/2, then from 3/2 to 1/2.

The + and − symbols in figure 4.18 refer to **parity**. Nuclear parity is like your right and left hands: they look the same, but they're really mirror images of each other. Likewise a nucleus with + (even) parity looks like a mirror image of an isomer with − (odd) parity. Parity also indicates the direction of the spin. Parity also plays a role in determining whether a particular isomeric transition can take place, but this is beyond the scope of this text. We will limit ourselves to observing which transitions do not take place, and rationalizing them simply on the basis of the magnitude of change in spin.

When the nucleus changes from one spin state to another, energy is released. We've already seen that it can be emitted in the form of a high-energy (gamma) photon. In addition to gamma emission, the atom can also spit out one of its electrons to release the energy. This is called **internal conversion**, and this is why this section is titled isomeric transitions rather than gamma decay. When a nucleus transforms from one isomer to another, it has two options to rid itself of the energy it just created.

The emitted electron is called a **conversion electron**. Conversion electrons usually come from the K shell (K int cov), as these electrons have the highest probability of being close to the nucleus. The kinetic energy of the conversion electron is simply the transition energy minus the electron-binding energy. The lowest energy excited state for ^{125}Te is 35.5 keV above the ground state. It undergoes isomeric transition to the ground state according to the probabilities given in table 4.3.

The transition energy (total energy of decay) is equal to the energy of the gamma photon (35.5 keV) because there are no nuclear excited states between the ground state and this one. The energies of the conversion electrons are all lower than the

TABLE 4.3

^{125}Te* Isomeric Transitions

Emission	Probability (%)	Energy (keV)
γ	6.66	35.5
K int cov	80.00	3.7
L int cov	11.42	31.0
M int cov	1.90	34.5

transition energy by an amount equal to the electron-binding energy for that shell. Using the numbers above we can calculate those energies.

K shell electron binding energy $= 35.5$ keV $- 3.7$ keV $= 31.8$ keV
L shell electron binding energy $= 35.5$ keV $- 31.0$ keV $= 4.5$ keV
M shell electron binding energy $= 35.5$ keV $- 34.5$ keV $= 1.0$ keV

The probabilities of each type of emission following the ^{125}Te* $\rightarrow$ ^{125}Te transition are also presented in table 4.3. Notice that internal conversion is much more likely than gamma emission. This is partially expected since Te has a moderately high atomic number ($Z = 52$). Also note that emission of a K shell electron is most probable, but occasionally electrons from the L and M shells are booted.

The transition used above is rather low in energy. If the transition energy is lower than the binding energy of the electrons, then internal conversion for that shell is not possible. Isomeric transitions that are of greater energy can emit fairly high-energy electrons. Once emitted, they look an awful lot like a beta particle. The only difference between conversion electrons and beta particles is their origin: conversion electrons come from the atom's orbital electrons, while beta particles are created in the nucleus.

The emission probabilities vary by element *and* by the particular transition. For example, there's another excited state for ^{125}Te that is 145 keV above the ground state. It has a half-life of 57.4 days, and is therefore a metastable state and can be designated as ^{125m}Te. ^{125m}Te has two options for decay: straight to the ground state, or to the other ^{125}Te*. Both of those transitions will have different probabilities for emission from those listed in table 4.3.

The relative probability of internal conversion to gamma emission is given by the **conversion coefficient** (α_T). It is simply the ratio of ejected electrons to gamma photons:

$$\alpha_T = \frac{I_e}{I_\gamma} \qquad (4.4)$$

Instead of counting electrons and photons, we can use the probabilities in table 4.3 to calculate the conversion coefficient for the ^{125}Te* $\rightarrow$ ^{125}Te transition:

$$\alpha_T = \frac{I_e}{I_\gamma} = \frac{80.00 + 11.42 + 1.90}{6.66} = 14.0$$

Additionally, we could calculate the conversion coefficients for the individual electron shells:

$$\alpha_K = \frac{I_e}{I_\gamma} = \frac{80.00}{6.66} = 12.0 \qquad \alpha_L = \frac{I_e}{I_\gamma} = \frac{11.42}{6.66} = 1.71 \qquad \alpha_M = \frac{I_e}{I_\gamma} = \frac{1.90}{6.66} = 0.285$$

where α_K is the conversion coefficient for internal conversion of K shell electrons, α_L is for the L shell, and α_M is for the M shell.

4.8   OTHER DECAY MODES

4.8.1   SPONTANEOUS FISSION

Spontaneous fission is when a nuclide splits into two large chunks. This is a significant mode of decay for some of the very heavy nuclides, and will be discussed in more detail in 8 chapter.

4.8.2   CLUSTER DECAY

Cluster decay happens when very heavy nuclides can also blow chunks bigger than an alpha particle, but not quite big enough to qualify for spontaneous fission. ^{12}C, ^{14}C, ^{16}O, inter alia, are (very rarely) emitted from large nuclides in a desperate attempt to quickly form a stable nuclide.

4.8.3   DOUBLE BETA DECAY

Double beta is a rather unusual form of decay, and typically occurs with a neutron-rich *ee* nuclide. An example is ^{82}Se:

$$^{82}_{34}\text{Se}_{48} \rightarrow\, ^{82}_{36}\text{Kr}_{46} + 2\,^{0}_{-1}e + 2\overline{\nu}$$

If ^{82}Se could decay via regular beta emission, it would produce $^{82}_{35}\text{Br}_{47}$, which is *oo*. Instead, ^{82}Se simultaneously emits two beta particles, decaying directly to ^{82}Kr. Two antineutrinos are also emitted to satisfy the conservation laws discussed earlier in this chapter.

4.8.4   DELAYED PARTICLE EMISSIONS

Delayed particle emissions are sometimes observed for nuclides at the extremes of the valley of stability, that is, up on a high ledge overlooking the valley. If a nuclide is particularly neutron- or proton-rich it can sometimes emit a particle *after* the decay. This occurs when the daughter is formed in a particularly high-energy excited state and is an alternative to an isomeric transition to a lower excited state or to the ground state. In this case an isobar is not formed as the product. For example, almost 7% of all ^{137}I decays emit a beta particle then a neutron (93% of the time it just emits a beta particle). The minor branch can be represented as:

$$^{137}_{53}\text{I} \rightarrow {}^{136}_{54}\text{Xe} + {}^{0}_{-1}e + {}^{1}_{0}n + \bar{\nu}$$

Although, a better representation would include the fleeting formation of the excited daughter:

$$^{137}_{53}\text{I} \rightarrow {}^{137}_{54}\text{Xe}^* + {}^{0}_{-1}e + \bar{\nu}$$

$$^{137}_{54}\text{Xe}^* \rightarrow {}^{136}_{54}\text{Xe}^* + {}^{1}_{0}n$$

QUESTIONS

1. Calculate the nuclidic mass of ^{240}U from the primary mode of decay for ^{244}Pu. Write out the decay reaction and use 4.589 MeV as the kinetic energy of the most energetic alpha particle possible for this decay. What is the recoil energy of ^{240}U?

2. Using information from the chart of the nuclides, draw a decay diagram for ^{268}Mt. Calculate the recoil energy of the daughter. For whom are the parent and daughter elements named?

3. The most common beta particles observed in the decay of ^{59}Fe have E_{max} values of 0.466 and 0.274 MeV. The two most prominent γ-rays have energies of 1099 and 1292 keV. If the total energy of decay is 1.565 MeV and the atomic mass of the daughter is 59.930791 u, what is the atomic mass of ^{59}Fe? Draw a decay diagram consistent with the information given here.

4. Explain why ^{37}Ar might be expected to undergo positron decay, but doesn't.

5. ^{54}Mn (53.940363 u) decays to ^{54}Cr (53.938885 u) via electron capture. In almost every decay a gamma photon with an energy of 834.8 keV is emitted. Write out the decay reaction, and draw a decay diagram using only the information given here. Give two possible reasons why a gamma is *not* seen in a very small percentage of these decays.

6. The binding energies for electrons in nuclide X are given in table 4.4. Calculate the energy of the $M_2 \rightarrow K$ X-ray, and the $M_2M_3M_5$ Auger electron. What is the energy of the KL_2L_3 Auger electron? What kind of Auger electron would have an energy of 12.315 keV?

7. ^{64}Cu undergoes all three forms of beta decay. Draw a decay diagram consistent with the information given here. The total energy of decay for β^- is 0.579 MeV, and for β^+ and EC it is 1.6751 MeV. The only beta particle observed has an E_{max} of 0.578 MeV and the only positron has an E_{max} of 0.651 MeV. A 1346 keV gamma photon is observed, but only in a small fraction of the electron capture decays.

TABLE 4.4
Electron-Binding Energies for Element X

	keV
M_5	3.332
M_4	3.491
M_3	4.046
M_2	4.831
M_1	5.182
L_3	16.300
L_2	19.693
L_1	20.472
K	109.651

8. A nuclide decays, giving off two different positrons and a single γ-ray of 677 keV. If the total energy of decay is 6.138 MeV, what are the likely energies of the two positrons emitted?
9. Calculate the energy resulting from the reaction of 0.01 mol of positrons with an equal number of electrons.
10. Complete the first four (empty) columns in table 4.5 without any additional information, and then check your work using the chart of the nuclides. Use the chart to fill in the final column. Finally, draw a simplified decay diagram for this set of isobars, showing only the major mode of decay between nuclides.

TABLE 4.5
Data for Some $A=25$ Isobars

Isobar	N	Z	N/Z	Mode of Decay	Atomic Mass (u)
$^{25}_{9}\text{F}$					
$^{25}_{10}\text{Ne}$					
$^{25}_{11}\text{Na}$					
$^{25}_{12}\text{Mg}$					
$^{25}_{13}\text{Al}$					
$^{25}_{14}\text{Si}$					

11. Explain the m in ^{60m}Co. What is meant by the fact that its I_e/I_γ ratio is 41?
12. Using the chart of the nuclides, draw a decay diagram for ^{130}Te.
13. What are the differences between conversion coefficient and fluorescent yield?

5 Interactions of Ionizing Radiation with Matter

The previous chapters have established a firm understanding of radioactive decay. This chapter now turns to the ramifications of ionizing radiation. Understanding how ionizing radiation interacts with matter is essential to understanding how it can be detected and shielded, and how it affects living systems. We'll start with a more detailed definition of ionizing radiation, and then look at the various ways energetic particles and photons can interact with matter.

5.1 IONIZING RADIATION

Ionizing radiation refers to radiation with enough energy to knock an electron loose from an atom. Alpha particles, beta particles, positrons, γ-rays, and X-rays are all forms of ionizing radiation. Remember that X-rays and γ-rays are located on the high energy end of the electromagnetic spectrum (Chapter 1) and that the difference between the two is how they are generated (Chapter 4). X-rays are generated by electrons decellerating or dropping from a higher shell to the K or L shell, while γ-rays result from nuclear transitions (Chapter 4).

Just below X-rays on the electromagnetic spectrum is UV light. The line between nonionizing UV and ionizing X- and γ-rays is somewhat fuzzy, but is generally drawn at 100 eV. A threshold of 100 eV for ionization may seem artificially high, especially since the electron-binding energy of the outermost electrons (valence) for most atoms is less than 15 eV and most chemical bonds (sharing valence electrons) have energies of 1–5 eV. It seems like it should require much less than 100 eV to knock electrons loose from atoms, but we know that UV is not ionizing. As we will see in this chapter, only a fraction of the photon's (or particle's) energy can be transferred to an atom, and 100 eV is necessary to consistently knock electrons loose from atoms.

5.2 CHARGED PARTICLES

Interactions between charged particles (e.g., alpha or beta particles) and matter can be thought of as the transfer of energy from the particle to the matter. As a particle travels through matter, energy is transferred from it until it has about the same energy as the surrounding matter. It might be strange to think of, but all matter has energy, unless it is at absolute zero! As the particle transfers its energy, it slows down—its energy is kinetic energy and kinetic energy is equal to one half times mass times velocity squared:

$$KE = \frac{1}{2}mv^2 \tag{5.1}$$

As kinetic energy decreases, then velocity must also decrease, since mass remains constant. These interactions can also be thought of as collisions between the energetic particles and atoms, although this is not strictly correct. Even so, collisions are often used to describe the transfer of energy from particle to atom. Perhaps it is because that is how we often observe energy transfer in our macroscopic world (e.g., a bowling ball hitting the pins). Please keep in mind that the subatomic world is a very different place. Solids and liquids appear to us to pretty well occupy their space, but on the atomic level they are mostly empty space. An alpha particle traveling through matter is a bit like coasting on a bicycle into a flock of small birds. Each impact on a bird causes the bike rider to slow down a little bit as kinetic energy is transferred from the bicyclist to a bird. If the bicycle and rider represent an alpha particle, then the birds represent electrons. Interactions with electrons are much more likely than an interaction with the nucleus, as the electrons occupy most of the space inside an atom. If the bike rider is especially unlucky, it will strike an object inside the flock with a mass greater than or equal to it (the nucleus). If we continue to let the bike and rider represent an alpha particle, then the mass of two SUVs would represent a gold nucleus (ouchies!).

An **elastic** collision is one where kinetic energy of the particle is cleanly transferred as kinetic energy to the electron. The total kinetic energy of the particle and the electron it hits remains the same. An **inelastic** collision is one where some of the particle's kinetic energy is converted to some other form of energy (such as an electron moving to a higher shell). Inelastic collisions are very common between a large charged particle (proton, alpha, etc.) and an electron. When the electron is promoted to a higher energy level the interaction is an **excitation**. As shown in figure 5.1a, this process is reversible—the electron will eventually (usually right away) de-excite, emitting electromagnetic radiation such as visible light or X-rays.

If the amount of energy transferred by the particle is greater than the electron's binding energy, the electron is removed from the atom (fig. 5.1b). This is called **ionization** because it results in the formation of a cation and a free electron (an **ion pair**). The creation of an ion pair by ionizing radiation is called a primary ionization. This distinguishes it from secondary ionizations, which can subsequently be caused by the freed electron. Because the freed electron has a significant amount of kinetic energy, it can cause its own ionizations. Such ionizations are called **secondary ionizations**, and the freed electron is often called a **secondary electron**. Secondary electrons are sometimes also referred to as **delta rays**, although this is somewhat misleading, as the term "rays" refers to photons, not particles. As indicated in figure 5.1b, ionizations

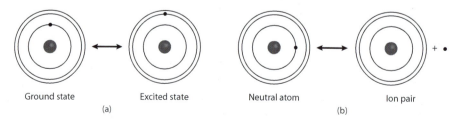

Ground state   Excited state   Neutral atom   Ion pair

(a)   (b)

FIGURE 5.1   (a) Electron excitation and (b) atomic ionization.

are also reversible, although it is rarely the same electron that returns to the cation reforming the neutral atom.

Excitations and ionizations are two very common results of the interaction of ionizing radiation with matter. What are the relative odds of ionizations and excitations? Electron excitations are roughly twice as common as ionizations in air.

Example: A 7.6868 MeV alpha particle emitted during the decay of ^{214}Po undergoes a collision with an M shell electron in an atom of holmium. If the alpha particle now has only 7.6830 MeV, will the electron be removed from the atom, and if so, how much kinetic energy will it have?

Remember from Chapter 4 the electron-binding energy is the amount of energy required to completely remove an electron from an atom. Thankfully, the energy of an M shell electron in holmium is given in figure 4.10 (1.71 keV). The energy of the electron booted out of Ho will be the energy transferred minus the binding energy. The energy transferred is:

$$7.6868 \text{ MeV} - 7.6830 \text{ MeV} = 0.0038 \text{ MeV} = 3.8 \text{ keV}$$

This is greater than the binding energy of an M shell electron in Ho, so the atom will be ionized. The energy of the freed (secondary) electron is:

$$3.8 \text{ keV} - 1.71 \text{ keV} = 2.1 \text{ keV}$$

Note that the electron has very little energy, because it appears that very little energy is transferred. In fact, the problem above represents the maximum amount of energy that could be transferred. Remember that the alpha particle is very massive (you and your bike) compared to an electron (a small bird). At most, an alpha particle can transfer only $\sim 0.05\%$ of its total energy to an electron.

Let's take a closer look at the relationship between kinetic energy and velocity. Earlier in this chapter the following facile formula was quoted:

$$KE = \frac{1}{2}mv^2$$

Applying this relationship to the alpha particle in the previous example *should* allow calculation of its velocity. A couple of conversion factors are necessary to get conventional velocity units. Kinetic energy needs to be in joules (J), remembering that

$$J = \frac{\text{kg} \cdot \text{m}^2}{\text{s}^2}$$

a joule is the same as a kilogram meter squared per second squared. Converting the energy of the alpha particle emitted by ^{214}Po from MeV to J:

$$7.6868 \text{ MeV} \times \frac{10^6 \text{ eV}}{\text{MeV}} \times \frac{1.602 \times 10^{-19} \text{ J}}{\text{eV}} = 1.231 \times 10^{-12} \text{ J}$$

Since energy is in joules, the mass of the alpha particle (electron mass is not included!) needs to be expressed in kilograms:

$$4.0015 \, u \times \frac{1.6605 \times 10^{-27} \, kg}{u} = 6.6445 \times 10^{-27} \, kg$$

Plugging energy and mass into $KE = 1/2mv^2$:

$$1.231 \times 10^{-12} \, J = \frac{1}{2}(6.6445 \times 10^{-27} \, kg) \times v^2$$

$$v = 1.925 \times 10^7 \, m/s$$

That looks like a pretty zippy little particle! The speed of light is 3.00×10^8 m/s, and no particle (matter) can travel at or above the speed of light. Our alpha particle isn't all that far from this ultimate speed limit. According to Einstein's Theory of Relativity, increasing amounts of energy are required to continue accelerating a particle whose velocity is approaching the speed of light. More and more energy is required for smaller and smaller increases in velocity, and the equation above ($KE = 1/2mv^2$—often referred to as the "classical equation") no longer holds true. In this case, it might be a good idea to calculate velocity using the more complex "relativistic equation" below:

$$KE = \left(\frac{1}{\sqrt{1 - v^2/c^2}} - 1 \right) mc^2 \tag{5.2}$$

where c is the speed of light. This equation is only useful for particles moving near speed of light. As written, it can be a little awkward to solve for velocity. Solving for velocity (not trivial!) we get:

$$v = c\sqrt{1 - \left(\frac{mc^2}{KE + mc^2} \right)^2} \tag{5.3}$$

If our alpha particle's velocity is any different using this equation, we'll know that relativity needs to be taken into account. If it ends up with the same velocity as the classical equation, we'll know that relativity doesn't apply here. Let's see!

$$v = 2.998 \times 10^8 \, m/s \times \sqrt{1 - \left(\frac{(6.6445 \times 10^{-27} \, kg) \times \left(2.998 \times 10^8 \, m/s\right)^2}{1.231 \times 10^{-12} \, J + (6.6445 \times 10^{-27} \, kg) \times \left(2.998 \times 10^8 \, m/s\right)^2} \right)^2}$$

$$v = 1.922 \times 10^7 \, m/s$$

Once again, it is important to use compatible units. Try the math yourself—there are a lot of buttons to press on this one and it's important to get them in the right order. The result for velocity is nearly the same as with the classical equation. Had we not carried everything out to at least four significant figures, we might not have seen any difference. Despite the small difference, using the relativistic equation is the correct approach for this problem. If the velocity were higher, or if the problem involved a less massive particle (like a proton or an electron), the difference between classical and relativistic equations would be more dramatic.

Velocities near the speed of light are often expressed as a fraction of the speed of light (v/c). For the alpha particle above this fraction would be:

$$\frac{v}{c} = \frac{1.922 \times 10^7 \text{ m/s}}{2.998 \times 10^8 \text{ m/s}} = 0.06412$$

Therefore the alpha particle could be said to be traveling at 6.412% of the speed of light. Despite our earlier amazement at its velocity, this is not really all that fast for a subatomic particle. Notice also that calculating v/c marginally simplifies the relativistic equation:

$$\frac{v}{c} = \sqrt{1 - \left(\frac{mc^2}{KE + mc^2}\right)^2} \tag{5.4}$$

The faster a particle is moving, the more energy it has to deposit in the matter it travels through. If the particle is traveling through air, an average of 33.85 eV is required for each ionization of an air molecule. Remember that air is ~78% nitrogen (N_2) and ~21% oxygen (O_2), so 33.85 eV/IP is reflective of this particular gas mixture. The average energy required to produce an ion pair in any medium is called its **W-quantity**. This energy cost includes the electron-binding energy, and is therefore the total energy lost by the ionizing radiation as it travels through the medium, and is not the energy of the secondary electron.

As a charged particle travels through matter, it will create a certain number of ion pairs per unit length. This is its **specific ionization** (SI), which depends on the energy of the particle and the nature (density) of the matter it is traveling through. The SI of alpha particles traveling through air typically varies from three to seven million ion pairs per meter (IP/m)—that's a lot of ions!

If we know how many ion pairs a charged particle creates per unit length (SI), and we know the average cost of creating each ion pair (W-quantity), we can also determine the average energy lost by the charged particle per unit length. This is called the particle's **linear energy transfer** (LET), and is obtained by multiplying the SI by the W-quantity:

$$LET = SI \times W \tag{5.5}$$

Example: Alpha particles of a certain energy have a SI of 5.05×10^6 IP/m. Calculate the LET value for these particles in air.

Since $W = 33.85$ eV/IP in air:

$$\text{LET} = \frac{5.05 \times 10^6 \text{ IP}}{\text{m}} \times \frac{33.85 \text{ eV}}{\text{IP}} = 1.71 \times 10^8 \text{ eV/m} = 171 \text{ MeV/m}$$

Wow! That's a lot of energy for every meter traveled in air. The alpha particle will need to be very energetic to travel a full meter, or it will only travel some fraction of a meter. **Range** is defined as the average distance a charged particle will travel before being stopped. For alpha and other "heavy," charged (nuclei) particles, range is calculated by dividing the energy of the alpha particle by its LET. This equation only applies to alpha particles and other nuclei:

$$\text{Range} = R = \frac{E}{\text{LET}} \tag{5.6}$$

Example: ^{241}Am is used in smoke detectors because the LET of air is significantly lower than the LET of smoky air. Calculate the LET and range of the 5.4857 MeV alpha particles emitted by ^{241}Am (SI $= 3.4 \times 10^4$ IP/cm).

$$\text{LET} = \frac{3.4 \times 10^4 \text{ IP}}{\text{cm}} \times \frac{33.85 \text{ eV}}{\text{IP}} \times \frac{\text{MeV}}{10^6 \text{ eV}} = 1.2 \text{ MeV/cm}$$

$$R = \frac{E}{\text{LET}} = \frac{5.4857 \text{ MeV}}{1.2 \text{ MeV/cm}} = 4.8 \text{ cm}$$

These alpha particles are completely stopped by only 4.8 cm of air! The smoke detector works by detecting the alpha particles after they've passed through $\sim$1–2 cm of air. If smoke gets between the ^{241}Am source and the detector, the LET increases; the detector does not see as many alpha particles per second and sounds the alarm. ^{241}Am is a pretty good source for this application as it has a 433-year half-life, so its activity is unlikely to change over the lifetime of the detector, yet it is short enough to only require a small amount ($\sim$1 µCi).

As mentioned above, SI varies with the energy of the article. As a particle moves through matter it loses energy, so we might expect the SI to change as a particle moves through matter. Figure 5.2 shows a representation of how SI is affected for a 7 MeV alpha particle moving through air.

Notice that the SI does not change much during the first few centimeters, and then it increases dramatically in the last fraction of a centimeter. This means that the alpha particle transfers energy to the air at a relatively constant rate, until it reaches some low energy threshold. After that point it begins to transfer energy to the air at ever-increasing rates until, suddenly, it is out of gas and stops.

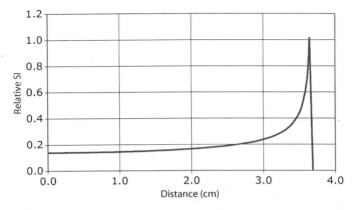

FIGURE 5.2 The change in specific ionization (SI) as a 7 MeV alpha particle penetrates air.

The behavior illustrated in figure 5.2 is common to all charged particles (alpha, beta, protons, …) in all media (air, water, humans, …). The peak observed in figure 5.2 is called the **Bragg peak** and represents the fact that charged particles deposit a large percentage of their energy at some depth inside the matter they penetrate. One way to think about this is in terms of the probability of interaction. While the particle is moving quickly (higher energy), it spends only a very short time by each atom it passes, decreasing the probability of interaction. At some point it is traveling slowly enough (lower energy) that the time spent near an atom increases the probability of interaction. From that point forward, the probability of interaction increases dramatically as the particle continues to decelerate. It's a bit like skipping a flat rock over a still lake. The first couple of skips are far apart, and relatively little energy is transferred per unit distance. When the rock slows, it starts to skip often over short distances, transferring a fair bit of energy over a relatively short distance before coming to a stop and sinking into the lake.

What happens when the particle stops? If it's an alpha particle it will steal a couple electrons and form a helium-4 atom; if it's a positron, it'll annihilate with an electron, etc. Keep in mind that the particle's energy does not go to zero, so it doesn't really "stop"—this is only possible at 0 K. The particle will have translational and rotational energy that reflect the temperature of the matter it is now part of. Instead of suggesting the particle comes to a stop, it is better to state that the particle transfers energy to its surroundings until its own energy matches that of its environment.

So far we've focused on the interactions of charged particles with air. What about their interactions with denser media, such as liquids and solids? Solids and liquids have a lot more matter packed into less space. Therefore, they have greater densities and the probability of interaction will increase dramatically. Specific ionization will increase (more energy deposited per centimeter) and range will decrease. One way to compare the shielding effectiveness of different materials is to compare relative stopping power (RSP):

$$\text{RSP} = \frac{R_{\text{air}}}{R_{\text{abs}}} \tag{5.7}$$

where R_{air} is the range in air and R_{abs} is the range in another *abs*orbing material. Larger values for RSP mean that that material is more strongly absorbing. Table 5.1 provides some RSP and range values for a 7.0 MeV alpha particle in different media.

As the density increases, so does RSP, and the range drops dramatically. The bottom line here is that alpha particles are not very penetrating—something we already knew to be true, but now we've got some serious numbers to back up the claim. Since we are made up mostly of water, the numbers above can be approximated for us. Skin averages between 1 and 2 mm thick (that's 0.1–0.2 cm), so we can feel pretty safe around alpha radiation. So long as it remains external to our bodies, the outer layers of our skin will stop alpha radiation.

Much of the discussion in this section has focused on "large" charged particles, such as alpha particles, protons and other nuclei. What about smaller charged particles like electrons? They could be beta particles, Auger or conversion electrons emitted during nuclear decay, or they could be an artificially produced electron beam. We should also consider positrons here as they have the same mass and unit charge. As a result, they tend to interact with matter in much the same way as electrons.

Like the heavier charged particles, electrons can cause excitations and ionizations as they pass through matter. Additionally, the probability of interaction increases as the velocity (KE) of the **incident electron** decreases. The incident electron is the electron that is transferring the radiation to the matter, that is, the radiation. The nature of the interaction and the energy of any secondary electrons produced from these interactions will depend on how much energy is transferred from the incident electron.

The obvious difference between electrons and larger, charged particles is mass. Electrons are thousands of times less massive than even the smallest nuclei. This has two major implications. First, if an electron and proton have the same energy, the electron will be moving a whole lot faster. Electrons will therefore require use of the relativistic kinetic energy equation at much lower energy values than protons or alpha particles. Since they are moving faster, they will spend less time passing atoms, and therefore have a lower probability of interaction. This means that electrons will

TABLE 5.1

RSP and Range Values for a 7.0 MeV Alpha Particle in Various Media

Absorber	Density (g/cm³)	RSP	Range (cm)
Air	0.0012	1	6.1
Water	1.0	970	0.0063
Aluminum	2.7	1700	0.0035

TABLE 5.2

Comparison of SI and Range Values for an Alpha Particle, a Proton, and an Electron

Particle	SI (air, IP/cm)	Range (air, cm)	Range (water, cm)
5.0 MeV alpha	40,000	3.6	0.0038
5.0 MeV proton	4200	35	0.036
5.0 MeV electron	60	1800	2

be more penetrating than alpha particles or protons with equivalent energy values. Lower probabilities of interaction mean fewer ion pairs created per centimeter and longer ranges, as illustrated in table 5.2.

Specific ionization values for electrons in air can be calculated if their velocity (energy!) is known, using Equation 5.8:

$$SI = \frac{4500 \text{ IP/m}}{v^2/c^2} \tag{5.8}$$

where v is the electron's velocity and c is the speed of light. Note that this equation works only for electrons and positrons in air.

Example: Calculate the SI and LET in air for the most probable energy beta particle emitted by ^{32}P.

Remember that beta particles share their energy with the antineutrino, and that the most probable energy is $0.3E_{max}$. For ^{32}P, $E_{max} = 1.709$ MeV, and therefore $0.3E_{max} = 0.51$ MeV. Using the relativistic equation, $v/c = 0.87$ for a 0.51 MeV electron (check it yourself!).

$$SI = \frac{4500 \text{ IP/m}}{(0.87)^2} = 6000 \text{ IP/m} = 60 \text{ IP/cm}$$

$$LET = SI \times W = \frac{6000 \text{ IP}}{m} \times \frac{33.85 \text{ eV}}{IP} \times \frac{MeV}{10^6 \text{ eV}} = 0.20 \text{ MeV/m}$$

Note that the same average energy per ion pair (W) is used here as was used for alpha particles. The source of energy doesn't matter; the average energy it takes to ionize air molecules remains the same. Note also that a higher energy beta particle would have a lower SI and LET. Higher energy means the particle is moving faster, has a lower probability for interaction, is more penetrating, and creates fewer ion pairs per meter.

The second major difference between electrons and nuclei as projectiles is that electrons are much more likely to be deflected (scattered) as they pass through matter. If most of the interactions between a charged particle and matter involve the electrons, this makes sense. If alpha particles are like a human cyclist coasting into a flock of birds, then electrons are simply a (high-speed) bird flying into a flock.

Both are slowed by collisions with the flock, but the cyclist's direction is unlikely to change much and the bird will likely bounce around like a cue ball slamming hard into a bunch of other billiard balls. Because of scattering, calculation of the average range for a particular energy beta particle is complex, and will not be discussed in this text.

Scattering of electrons by incident electrons doesn't really change the interactions from the perspective of the matter the particle is traveling through. These electrons will still experience excitations or ionizations as energy is transferred from the incident electron to the matter. The big difference is that electron projectiles can also be scattered by flying near a nucleus. The most important such interaction is called **inelastic scattering by nuclei**. If the projectile electron passes close enough to a nucleus, it will feel an attractive (positrons will be repelled) Coulomb force. The electron slows and curves around the nucleus, as shown in figure 5.3.

Because the electron slows down (loses kinetic energy) and energy needs to be conserved, the lost energy is emitted as a photon (electromagnetic radiation—the squiggly line in figure 5.3), typically an X-ray. The level of interaction will vary depending on how close the electron travels to the nucleus, and how many protons (Z) are in the nucleus. If the electron passes very close to the nucleus, it will be scattered through a higher angle, slow more, and emit a higher energy photon than if it were to pass further away. An infinite number of these kinds of interaction are possible, as the distance between the electron and nucleus can be varied by infinitesimally small increments. The result is that a wide variety of scattering angles can occur corresponding to a wide variety of energies for the emitted X-rays. An energy spectrum of X-rays produced from the beta particles emitted during the decay of ^{90}Sr and ^{90}Y interacting with aluminum is illustrated in figure 5.4.

X-rays produced by inelastic scattering of electrons by nuclei are often referred to as **bremsstrahlung**, which is a German word roughly translated as "braking radiation." Because they are a broad energy distribution, they are also referred to as **continuous X-rays**. Notice in figure 5.4 that not all energies have equal probabilities. In general the probability decreases as the energy increases. This is because lower-energy X-rays are produced when the electron is further from the nucleus, and there are more points in space at a greater distance. Prove this to yourself by drawing two concentric circles. If the nucleus is at the center and each point on the larger circle represents the probability of the lower energy interaction, and each point on the

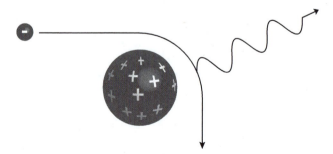

FIGURE 5.3   Inelastic scattering of an electron by a nucleus.

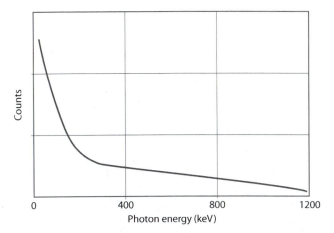

FIGURE 5.4 X-ray spectrum produced by beta emissions from ^{90}Sr and ^{90}Y interacting with Al.

smaller circle represents the probability of the higher energy interaction, which is more likely?

The effect of increasing the atomic number (Z) of the matter the electron is traveling through can be thought of in two ways. As mentioned above, the intensity of the interaction will increase with Z at the same distance from the nucleus. This is simply due to the increased positive charge of the nucleus. The other effect of increasing Z is that the nucleus will have greater reach. A larger positive charge means that the electron can interact with it from further away. Overall, as Z increases, the probability of inelastic scattering also increases.

Elastic scattering by nuclei happens when the projectile electron hits the nucleus head on and kinetic energy is conserved. This type of scattering is much less probable than inelastic scattering. The nucleus is small, and Coulomb repulsion can act over reasonably long distances. As illustrated in figure 5.5a, we can think of the electron initially in motion, heading right toward a nucleus, which is not moving. The collision itself is too violent to be depicted here, but figure 5.5b shows the aftermath. The electron has reversed direction and has lost some velocity, and the nucleus is now moving very slowly in the opposite direction. Only some of the electron's kinetic energy is transferred to the nucleus and, because the nucleus is so much more massive, it has very little velocity.

Elastic scattering of electrons by nuclei is also known as **backscattering** because the electron ends up going back along the path it came from.

Cerenkov radiation is an obscure form of radiation that results from particles (typically electrons) traveling from one medium into another. The speed of light depends (slightly!) on the medium it is traveling though. For example, light travels $1.33 \times$ slower in water than in a vacuum. This means that beta particles traveling near the speed of light in one medium that pass into another medium, where light travels more slowly, could now be exceeding the ultimate speed limit. Since no object can move faster than light, the electron has to slow down. In doing so, it gives off photons in the form of a bluish light. This effect isn't really important, but

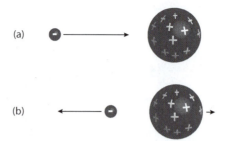

FIGURE 5.5   Elastic scattering of an electron by a nucleus.

it explains why nuclear reactor cores (Chapter 9) have a cool blue glow. High-energy beta particles are emitted by some of the products formed by the fission reactions and travel from the solid fuel pellet, through the solid fuel rod, and into the liquid (usually water) coolant.

5.3   PHOTONS

The transfer of energy from X- and γ-ray photons to the matter they traverse is a bit more complicated than that from particles. Nine different interactions are possible, but we will only look at the three that are generally the most probable, and have the most significance for medical applications: Compton scattering, the photoelectric effect, and pair production. These three interactions all transfer energy from high-energy photons to matter through fairly complex mechanisms. We'll discuss them in fairly simple terms, but please keep in mind that they are not as simple as they are described here.

5.3.1   COMPTON SCATTERING

Compton scattering can be imagined as a collision between an incident (incoming) photon and an outer-shell (valence) electron (fig. 5.6). In reality, it is better to think of it as a net transfer of some energy from the incident photon to a valence electron. There's not much electron-binding energy to overcome here (valance electron), so the electron gets booted away from its atom. The ejected electron is called the Compton (or recoil)[1] electron. Energy is conserved in this process, therefore the energy of the incident photon (E_0) equals the energy of the scattered photon (E_{SC}) plus the energy of the Compton electron (E_{CE}) plus the binding energy of the electron (BE):

$$E_0 = E_{SC} + E_{CE} + BE$$

Since the binding energy of a valence electron is really small compared to the other energies, we can neglect it. So the equation above simplifies to:

$$E_0 = E_{SC} + E_{CE} \qquad (5.9)$$

---

[1] An unfortunate term, as it has nothing to do with daughter recoil—the momentum of the daughter following nuclear (particularly alpha) decay.

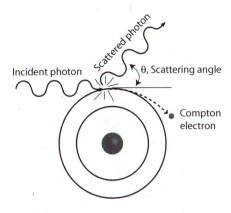

FIGURE 5.6   Compton scattering.

The photon is scattered, meaning its direction has changed. The angle through which the photon is scattered is called the scattering angle (θ). The energy of the scattered photon is related to its scattering angle and its original energy:

$$E_{SC} = \frac{E_0}{1 + \left(E_0 / 0.511\right) \times (1 - \cos\theta)} \tag{5.10}$$

All energies in the above equation *must* be in units of million electron volts. Basically, this equation tells us that as scattering angle increases, the energy of the scattered photon (E_{SC}) decreases. This makes some sense—if the incident photon "hits" the valence electron dead center, maximum energy is transferred and the scattered photon returns along the same path of the incident photon, traveling in the opposite direction (**backscattering**). Anything other than dead center will transfer less energy, and will deflect the photon less. At a minimum, a glancing blow will give a very small scattering angle and transfer only a tiny amount of energy.

   Example: Calculate the energies of the 140 keV γ photons that are produced during the decay of ^{99m}Tc when they are scattered through angles of 25° and 180°. Also determine the energies of the Compton electrons that are produced.

$$E_{SC} = \frac{0.140 \text{ MeV}}{1 + \left(\dfrac{0.140 \text{ MeV}}{0.511}\right) \times (1 - \cos 25°)} = 0.136 \text{ MeV or 136 keV}$$

$$E_{CE} = 140 \text{ keV} - 136 \text{ keV} = 4 \text{ keV}$$

$$E_{SC} = \frac{0.140 \text{ MeV}}{1 + \left(\dfrac{0.140 \text{ MeV}}{0.511}\right) \times (1 - \cos 180°)} = 0.0904 \text{ MeV or 90.4 keV}$$

$$E_{CE} = 140 \text{ keV} - 90.4 \text{ keV} = 50 \text{ keV}$$

TABLE 5.3

Energies of Scattered Photons and Compton Electrons when $\theta = 180°$

Nuclide	E_0 (keV)	E_{SC} (keV)	E_{CE} (keV)	% Transferred to Electron
^{125}I	27.5	24.8	2.7	9.8
^{99m}Tc	140	90.4	50	36
^{60}Co	1330	214	1116	84

As illustrated in table 5.3, the percentage of energy transferred from the incident photon to the Compton electron increases with the energy of the photon. All calculated values are for $\theta = 180°$ (backscattering). Higher-energy photons transfer a higher percentage of their energy to the Compton electron. Note also that a photon can never transfer all of its energy to an electron via Compton scattering. The maximum amount is transferred when the photon undergoes backscattering. In the ^{99m}Tc example above, it is when the photon transfers 50 keV to the Compton electron but retains 90.4 keV.

5.3.2 The Photoelectric Effect

When a high-energy photon interacts with an atom via the **photoelectric effect**, we can imagine that the energy of the incident photon is completely absorbed by an *inner*-shell electron (fig. 5.7). So long as the incident photon has enough energy ($>$binding energy), odds are good it will be a K shell electron that gets booted. This time the ejected electron is called a **photoelectron** and there is no scattered photon. The energy of the incident photon equals the energy of the photoelectron plus the binding energy. Since the binding energy of a K shell electron can be significant compared to the energy of a gamma photon, we can't neglect it.

$$E_0 = E_{PE} + BE \tag{5.11}$$

Kicking out an inner-shell electron creates an electron vacancy or hole in that shell. This is exactly the same situation that follows electron capture decay (Chapter 4). An electron in a higher-energy shell drops down to fill that vacancy and a characteristic X-ray or an Auger electron is then emitted.

5.3.3 Pair Production

The least important of the three photon–matter interactions we will study is **pair production**. Like the photoelectric effect, the atom absorbs the entire energy of the incident photon. The difference is that a β^- and a β^+ are created (fig. 5.8). They are the "pair" that is produced. The incident photon needs to be at least 1.022 MeV in order for pair production to take place. This is the amount of energy needed to create the amount of matter contained in an electron and a positron. If the incident photon

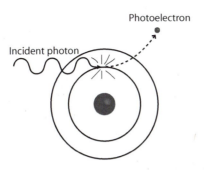

FIGURE 5.7 The photoelectric effect.

has more than 1.022 MeV of energy, the remainder goes into the kinetic energy of the electron and positron. In this case, the energy of the incident photon is equal to the energy of the emitted particles plus 1.022 MeV.

$$E_0 = E_{\beta+} + E_{\beta-} + 1.022 \text{ MeV} \qquad (5.12)$$

This extra energy is not evenly divided between the pair—anywhere from 20% to 80% of the excess energy can be deposited in either particle. The remainder goes into the other one.

The positron will transfer its kinetic energy to the nearby matter, likely via ionizations and excitations. When it runs out of gas, it will annihilate with an electron to form two 0.511 MeV γ-rays (fig. 5.8). Note that, theoretically, we could start with a 1.022 MeV incident photon and end up with two 0.511 MeV photons!

Of the three photon–matter interactions discussed here, pair production is the least important, because it is the least commonly observed for incident photons with low to moderate energies. So what are the relative probabilities of these three interactions? The relative probability of each depends on both the energy of the photon and the atomic number (Z) of the atom it interacts with.

Figure 5.9 illustrates how the relative probabilities of Compton scattering (solid black lines), the photoelectric effect (dashed lines), and pair production (solid gray line) vary with photon energy. Photon energy increases from left to right, and probability increases from bottom to top on each graph. Graph (a) is for water, (b) for Fe, and (c) is for Pb. For water (fig. 5.9a), Compton scattering is most likely for all photon energies graphed. For Fe (fig. 5.9b), Compton scattering still dominates, but the odds of the photoelectric effect for low-energy photons, and the odds of pair production at higher energies, are both increased relative to water. This trend continues as we move further down the periodic table. The photoelectric effect is dominant for low-energy photons traveling through lead (fig. 5.9c) and pair production becomes the most significant interaction for high-energy photons.

Notice also that the probability (y-axis) scale increases dramatically with atomic number for the three graphs in figure 5.9. It is generally true that the overall probability of photon interaction with matter increases as the atomic number increases. This can be understood in terms of density. The more matter you pack into the same amount

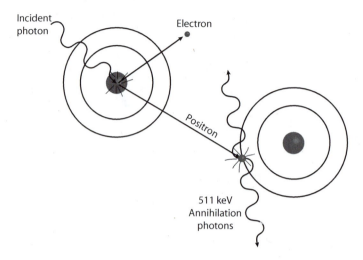

FIGURE 5.8   Pair production.

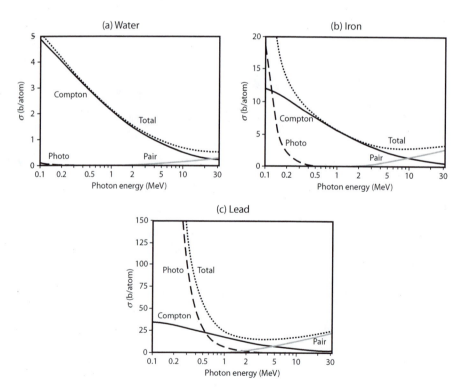

FIGURE 5.9   Relative probabilities of photon–matter interactions with (a) H_2O, (b) Fe, and (c) Pb. Produced using data from XCOM http://physics.nist.gov/.

of space, the more likely you'll get high-energy photons to interact within that space. Density generally increases with atomic number.

Finally, it is worth noting that the overall probability of interaction generally decreases with increasing energy of the photon. The overall probability that a photon will interact is the sum of the probabilities for Compton scattering, photoelectric effect, and pair production. For every element in figure 5.9 (dotted lines), the overall probability of interaction generally decreases as the energy of the photon increases. This is normally true of photon–matter interactions.

5.4   ATTENUATION OF X- AND γ-RADIATION

When photons run through matter, three things can happen. They can be absorbed (photoelectric effect or pair production), scattered (Compton scattering), or they can travel straight through. If a photon is scattered or absorbed, we say it has been attenuated. For example, if 1000 photons hit a slab of matter (a half-pound cheeseburger?) and 200 are scattered, 100 absorbed and 700 transmitted, then we'd say that 300 were attenuated.

What if a photon in the middle of a broad beam is scattered through a small angle? Would we notice? Only if our detector is far enough away. What if the beam is made up of photons of various energies (polyenergetic)? As we can see in figure 5.9, low-energy photons are always more likely to interact with matter than their high-energy buddies. It can get complicated.

Let's keep it simple for now, and assume that we're working only with a narrow beam of monoenergetic (all having the same energy) photons, all traveling in the same direction. This situation is often referred to as "good geometry," and anything else would be "bad geometry." When we have good geometry, the following equation holds:

$$I = I_0 e^{-\mu x} \tag{5.13}$$

which can be rewritten as:

$$\ln \frac{I}{I_0} = -\mu x \tag{5.14}$$

where I is the number of photons making it through the matter (transmitted), I_0 is the original number of photons in the beam, x is the thickness of the matter, and μ is the linear attenuation coefficient. The ratio of I/I_0 is also known as the **transmission factor**—which makes sense, as it is the fraction of photons that make it straight through the matter. Conversely, the attenuation coefficient is a measure of how much scattering and absorption take place, that is, it is based on how many photons don't make it straight through. It depends on the material and the energy of the photon beam. It should always be the same for a particular material and a specific photon energy. The units on μ are usually per centimeter cm^{-1}, requiring x to have units of centimeters. Other units of length can be used, but should cancel when used in the same calculation. With this equation, it should be apparent that we could never fully attenuate a photon beam. There is always a finite probability that some photons in any beam will travel through any matter.

Example: A narrow beam of 2000 monoenergetic photons is reduced to 1000 photons by a 1.0 cm thick piece of copper foil. What is the attenuation coefficient of the Cu foil?

$$\ln\frac{1000}{2000} = -\mu \times 1.0 \text{ cm}$$

$$\mu = 0.69 \text{ cm}^{-1}$$

In this example, the beam is attenuated to half its original intensity; therefore 1.0 cm is the **half-value layer** (HVL)—sometimes called the half value thickness (HVT). Since the HVL will always have a I/I_0 of 1/2, HVL can be mathematically defined as:

$$HVL = \frac{\ln 2}{\mu} \tag{5.15}$$

Some HVL values in water for various photon energies are listed in table 5.4. Notice that HVL increases with the photon energy. As the photon's energy increases, it takes more water to attenuate half of the beam, and the photons become more penetrating.

Example: What would the beam intensity be in the previous example if the foil's thickness is tripled?

$$I = I_0 e^{-\mu x} = 2000 \text{ photons} \times e^{-0.69 \text{ cm}^{-1} \times 3.0 \text{ cm}} = 250 \text{ photons}$$

This makes sense! For every HVL, half of the beam is cut out. We now have three HVLs in place. The first centimeter cuts the beam from 2000 to 1000, the second

---

TABLE 5.4

Half-Value Layers in Water for Different Photon Energies

Photon Energy (MeV)	HVL in Water (cm)
0.10	4.0
0.50	7.2
1.0	9.8
5.0	23
10	31
20	38

from 1000 to 500, and the third from 500 to 250. This is Exponential change, just like radioactive decay! A **tenth-value layer** (TVL) is the thickness required to attenuate a photon beam to one-tenth of its original value.

$$TVL = \frac{\ln 10}{\mu} \tag{5.16}$$

The attenuation coefficient can take other forms and units. Some look pretty crazy. The mass attenuation coefficient (μ_m) is:

$$\mu_m = \frac{\mu}{\rho} \tag{5.17}$$

where ρ is the density of the absorbing material. If density has units of grams per cubic centimeter and μ is per centimeter, then μ_m would have units of square centimeters per gram.

The atomic attenuation coefficient (μ_a) is:

$$\mu_a = \frac{\mu \times \text{atomic mass}}{\rho \times \text{Avogadro's number}} \tag{5.18}$$

Believe it or not, μ_a has units of square centimeters per atom.

Example: Calculate the mass and atomic attenuation coefficients for the copper example above.

The density of copper is 8.96 g/cm^3 and its average atomic mass is 63.55 g/mol.

$$\mu_m = \frac{0.69 \text{ cm}^{-1}}{8.96 \text{ g/cm}^3} = 0.077 \text{ cm}^2/\text{g}$$

$$\mu_a = \frac{0.69 \text{ cm}^{-1} \times 63.55 \text{ g/mol}}{8.96 \text{ g/cm}^3 \times \left(6.02 \times 10^{23} \text{ atom/mol}\right)} = 8.2 \times 10^{-24} \text{ cm}^2/\text{atom}$$

Remember, all of this is for good geometry (monoenergetic, narrow beam). With a broad beam, percentage transmittance doesn't drop off as rapidly, because some photons can be scattered through small angles, and they might not escape the beam by the time we detect them.

Polyenergetic beams are also problematic because μ varies with the energy of the photon. As noted for figure 5.4, low-energy photons are more likely to interact. Therefore, a polyenergetic beam will have a higher energy distribution after it passes through attenuating matter. This is called **filtration**. Higher-energy photons are called "hard," and lower-energy photons "soft." A beam that has been filtered would then be "harder." Filtration is commonly done with medical X-rays. Most medical

X-ray machines produce a polyenergetic beam that is filtered before it hits the patient (usually with an aluminum disk—take a look the next time you go to the dentist!). This is a good thing, as the soft X-rays are more likely to be absorbed by the patient, unnecessarily increasing their dose during the procedure.

For polyenergetic beams, an effective attenuation coefficient can be determined:

$$\mu_{eff} = \frac{\ln(I_0/I)}{x} = \frac{\ln 2}{HVL} \tag{5.19}$$

It is based on measurements, and is specific to the X-ray source, as different sources will produce different spectra. Calculation of other flavors of the attenuation coefficient is performed in the same way as before.

Example: An X-ray beam has an HVL of 1.5 mm for Cu. What are the effective linear and mass attenuation coefficients?

$$\mu_{eff} = \frac{\ln 2}{0.15 \text{ cm}} = 4.6 \text{ cm}^{-1}$$

$$\mu_{meff} = \frac{\mu_{eff}}{\rho} = \frac{4.6 \text{ cm}^{-1}}{8.96 \text{ g/cm}^3} = 0.52 \text{ cm}^2/\text{g}$$

QUESTIONS

1. Calculate the wavelength and frequency of a photon with 100 eV of energy.
2. An alpha particle emitted during the decay of ^{222}Rn (5.4895 MeV) transfers 0.0010% of its energy to an argon atom. Is this enough energy to ionize the argon atom by booting any of its electrons? Table 5.5 provides electron-binding energies for all occupied subshells in an argon atom. If it can eject electrons from any of the shells listed below, calculate the energy of the secondary electron that is produced.
3. How many ion pairs will be formed by a single 5.00 MeV alpha particle traveling through air? Assume all of the particle's energy goes into ion-pair formation.

TABLE 5.5

Electron-Binding Energies for Argon

K	L_1	L_2	L_3	M_1	M_2	M_3
3.20 keV	327 eV	250 eV	248 eV	27 eV	14 eV	14 eV

4. If the range of a charged particle in dry air (at standard temperature and pressure) is known, what other data must be obtained to calculate the range of the same particle in titanium?

5. Calculate v/c for a 7.6868 MeV proton. Repeat for an electron (beta particle) of the same energy. Use both the classical and relativistic equations for both. Which gives more accurate results?

6. Calculate the kinetic energy (MeV) of an electron traveling at a velocity exactly equal to one half the speed of light. Repeat the calculation for a proton, and an alpha particle.

7. What is the maximum velocity for a beta particle emitted during the decay of ^{6}He? Why must this be considered a "maximum" velocity?

8. Calculate the mass of 1.0 μCi of ^{241}Am. AmO_2 is its chemical form in smoke detectors. What is its specific activity? Do ^{241}Am-containing smoke detectors pose a radiation hazard for people who have them in their homes?

9. Former KGB agent Alexander Litvinenko was poisoned with ^{210}Po in 2006. ^{210}Po emits an alpha particle with an energy of 5.3044 MeV and a range in air of 4.0 cm. What is the SI of this alpha particle in air? On average, how many ion pairs will be created in air by each alpha particle emitted by ^{210}Po? Why is ^{210}Po a particularly clever poison?

10. How would you expect protons to interact with matter? Compare to alpha particles and electrons of equal energy.

11. How many times more massive are alpha particles compared to electrons?

12. An elastic collision takes place between an 2.51 MeV electron and a stationary ^{12}C nucleus. Assuming that exactly 25% of the electron's energy is transferred to the ^{12}C nucleus, calculate the resulting velocities of both.

13. An electron initially traveling at 2.997×10^8 m/s slows as it transfers to another medium, emitting a photon of blue light (wavelength = 455 nm). Calculate the electron's new velocity. What is the blue light known as?

14. Calculate the energies of the backscatter peak and the Compton electron for a 662 keV γ-ray.

15. Calculate the energies of the scattered photon and the Compton electron when incident γ-radiation of 167 keV (from ^{201}Tl) is scattered through an angle of 23°. Assume the Compton electron originated from the L_1 shell of oxygen (binding energy = 37.3 eV).

16. What is the LET of beta particles emitted by ^{90}Y in air? Use the most probable energy of a beta particle emitted by ^{90}Y. Estimate the range in water of an alpha particle with the same energy, and compare to the average range of the ^{90}Y beta particle (11 mm) in human tissue.

17. Humans can be considered as mostly water. If a 25 cm thick human is exposed to gamma radiation from ^{137}Cs, estimate the percentage of the γ-ray intensity attenuated. The mass attenuation coefficient for water exposed to ^{137}Cs is 0.088 cm^2/g. What is the most likely interaction between photon and water molecule?

18. Copper has a density of 8.9 g/cm^3, and its total atomic attenuation coefficient is 8.8×10^{-24} cm^2/atom for 500 keV photons. What thickness of

copper is required to attenuate a 500 keV photon beam to half of its original intensity?

19. The mass attenuation coefficient for iron exposed to 1.5 MeV X-rays is 0.047 cm^2/g. Calculate the half- and TVLs.

20. Using information from the problem above, calculate the atomic attenuation coefficient. Also calculate the intensity of the original beam if the measured intensity is 5800 photons/s on the other side of a 70 mm thick piece of iron.

21. Calculate the kinetic energy of a photoelectron released from the K shell (binding energy $= 36.0$ keV) as a photon with a frequency of 9.66×10^{18} s^{-1} interacts with a chunk of cesium. What part of the electromagnetic spectrum does this photon belong to?

6 Detection of Ionizing Radiation

Perhaps the main reason that most people fear ionizing radiation is that humans have no way to directly sense it. We need instruments to convert ionizing radiation into signals that we can observe and understand. Remember that ionizing radiation can interact with matter by ionization (creation of an ion pair) or excitation. We can take advantage of both in our quest to detect ionizing radiation. This chapter is all about how these instruments work. Special emphasis will be placed on Geiger-Müller (GM) tubes (aka Geiger counters), as they are so commonly used, and on gamma spectroscopy, because of its importance to nuclear medicine.

6.1 GAS-FILLED DETECTORS

The basic design and function of any gas-filled detector is illustrated in figure 6.1. Basically, ionizing radiation passes through some gas in a container, creating ion pairs from the gas molecules. The negatively charged electrons then begin to migrate toward a positively charged piece of metal (anode) and the positively charged cations move toward a negatively charged cathode. When the ions hit the metal, they create an electrical signal (current), which can be read. Many different electrode configurations are possible, but the most common is a wire anode running down the center of a cylindrical cathode.

Remember from Chapter 1 that X- and γ-rays have a relatively low probability of interacting with matter. Since gases have very low densities, gas-filled detectors are quite inefficient in detecting photons. Typically <1% of all photons entering the tube will interact with the gas inside the tube. That means that >99% fly right on through, undetected. However, gas-filled detectors are very efficient at detecting beta and alpha radiation as they have much higher probabilities of interacting with matter.

The major difference between the three flavors of gas-filled detectors discussed below is the voltage applied to the electrodes. We will look at them in order of increasing voltage.

6.1.1 IONIZATION CHAMBERS

Ionization chambers are the lowest voltage form of gas-filled detector, typically running at 50–300 V. The idea is to have just enough voltage to keep the ions moving toward the electrodes without accelerating them. If they are accelerated, there's a chance they may cause other ionizations as they travel through the chamber. If we can keep their velocity more or less constant, then the only thing detected are the ions created.

Ionization chambers are usually filled with air. Remember that, on average, it only takes about 34 eV to create an ion pair in air; so many ion pairs can be expected

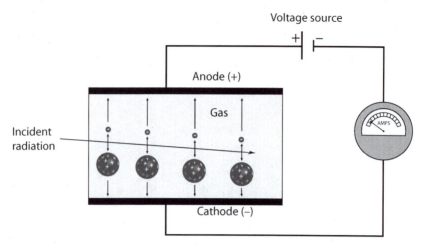

FIGURE 6.1   Schematic of a gas-filled detector.

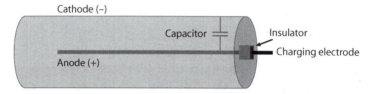

FIGURE 6.2   Schematic drawing of a pocket dosimeter.

from every particle or photon of ionizing radiation that enters the chamber. Even so, ionization chambers cannot always detect individual events, because the signal from these events can be too weak. Fortunately, most radioactive sources emit radiation at a pretty high rate, and the chamber can measure the overall current created, which can be calibrated to a count rate.

One application of ionization chambers is the pocket dosimeter. A schematic of one is pictured in figure 6.2. The inner wire (anode) and the outer case (cathode) are given a charge, which is stored in a capacitor. The user can then look through a window on one end and read the level of charge on a gauge. As ionizing radiation interacts with the gas inside the dosimeter, the charge is gradually dissipated. The extent of charge dissipation is a measure of the amount of radiation the user was exposed to. These dosimeters are commonly used when the possibility of a significant dose in a relatively short time exists. The accuracy of these devices is ~20%.

6.1.2   PROPORTIONAL COUNTERS

One disadvantage of ionization chambers is that they are not very sensitive. They are designed to detect radiation in bulk, not as individual photons and particles. They cannot detect a single ionizing particle because the signal generated is sometimes too weak. Fortunately we can make a couple of simple changes to amplify the signal.

First, we can crank up the voltage. A higher potential means that the electrons will now be accelerated as they approach the anode. If they accelerate enough, they can begin to cause additional ionizations, which will generate more electrons, which then accelerate, causing more ionizations... and when all these electrons hit the anode, a stronger signal is generated. This process of accelerating the electrons to create more is called a **cascade** or **avalanche**. Typically 10^6 times as many ion pairs are generated in the cascade as were originally generated by the ionizing radiation.

Secondly, we can change the gas. Air contains oxygen, which has a bad tendency to react with free electrons, preventing them from getting to the anode, dampening the signal. **Proportional counters** typically use Ar or Xe gas inside the detector, and run at higher voltages than ionization chambers.

The signal is now strong enough that individual counts can be made. Another bonus is that the total charge produced is *proportional* to the energy deposited by the ionizing radiation. Therefore, both count rate and energy information on the incident radiation can be obtained. This can be helpful in discerning alpha from beta radiation.

6.1.3   GEIGER-MÜLLER TUBES

Geiger-Müller (GM) tubes (fig. 6.3) are designed for maximum signal amplification. To do this the voltage is usually cranked up to around 800 V. The detector gas is almost always Ar. As a result, electrons are accelerated like crazy toward the anode. Typically 10^{10} times more ion pairs are created.

Lots of excitations also happen in the mad electron rush of electrons to the anode. When these atoms de-excite, electromagnetic radiation (visible light, UV, and X-rays) is given off—*lots* of it. Some of these photons have enough energy to knock electrons loose from the metal electrodes (Einstein's photoelectric effect). Some of the photons also cause additional ionizations in the detector gas.

This cascade could easily get out of hand, so what shuts it down? The slow moving cations! The cations are much more massive than the electrons, and therefore move much more slowly under the same conditions. So many ionizations are created near the anode by all the electrons accelerating toward it, that a cylindrical tube of cations forms around the anode. Eventually, all gas inside this tube is ionized, and the cascade shuts down.

As the cations finally approach the cathode ($\sim$300 μs later), electrons come off the wall to combine with them. As these electrons accelerate toward the vast cation cloud, they excite electrons in the Ar atoms they meet along the way, and more

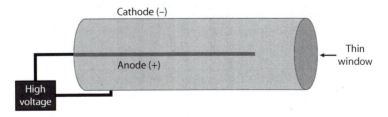

FIGURE 6.3   Schematic of a Geiger-Müller detector.

electromagnetic radiation is generated. This new burst of photons threatens to start another cascade. To shut this down (**quench**), chlorine is added to the Ar gas inside the GM tube:

$$Cl_2 \xrightarrow{\text{photons}} 2Cl\bullet \rightarrow Cl_2$$

Chlorine efficiently absorbs the higher-energy photons, which symmetrically break the Cl−Cl bond, forming two chlorine radicals. These radicals then recombine, dissipating the released energy as heat.

Because of the huge gas amplification (acceleration toward the electrodes), GM tubes detect individual events quite well. However, the strength of the observed signal is constant, regardless of the amount of energy deposited in the tube by the ionizing radiation. Therefore, only count information can be recorded with GM tubes, not energy.

For a certain amount of time after ionization takes place in a GM tube it is unable to distinguish another ionization of the Ar gas. Even after the cascade has been quenched, a little more time must pass before we can be sure another event will be recorded. A number of terms are applied to this time interval:

Dead time is the time interval after a pulse has occurred during which the counter is insensitive to further ionizing events.

Resolving time is the minimum time interval by which two pulses must be separated to be detected as two individual pulses.

Recovery time is the time interval that must elapse after a pulse has occurred before another full-sized pulse can occur.

The differences between these three terms are fairly subtle, but the implication of this required delay between counts is clear. At higher count rates, GM tubes will begin to miss some counts. For example, if two beta particles enter a GM tube at about the same time (**coincidence**), only one count will be recorded. What kind of a time interval are we talking about here? It depends a bit on the tube, but typically it is a few hundred microseconds. Because the possibility of two ionizations occurring within this dead time increases with the count rate, higher count rates mean more counts will be missed!

Fortunately, it is easy to estimate the true number of counts by experimentally determining how many counts are being missed at various count rates, and plotting the results in a graph such as that shown in figure 6.4. Correcting for the missing counts is usually called a **coincidence correction**. The straight line shows the entirely theoretical case when true counts are always equal to observed counts. The curved line shows the reality of observed counts being less than true counts at higher count rates. In the real world, count rates can also be affected by variances in line voltage, background, sample geometry (how the sample is positioned relative to the detector), atmospheric conditions, movement of wires, etc. As mentioned in Chapter 2, the percentage of decays detected relative to all of the source's decays in the same time interval is called the percent efficiency. As is clear here, the percent efficiency can be affected by a number of factors, and would therefore be expected to vary somewhat from detector to detector.

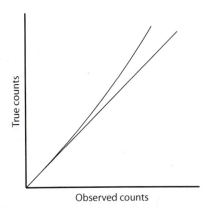

FIGURE 6.4 Coincidence correction graph for a GM tube.

6.2 SCINTILLATION DETECTORS

6.2.1 PHOTOMULTIPLIER TUBES

When ion pairs recombine, or excited electrons drop back to the ground state, visible light can be given off. **Scintillators** are materials that do this particularly well. Only a few thousand photons are usually produced when ionizing radiation travels through a scintillator. This is not very bright and, therefore, very difficult to detect, so we will need a way to amplify the visible light signal. **Photomultiplier tubes** (PMT) do this by converting the visible light photons to electrons then amplifying the signal. This process is illustrated in figure 6.5.

When the visible light photons leave the scintillator they hit the photocathode. The photocathode converts the photons into electrons (photoelectrons). These electrons now represent the signal generated by the ionizing radiation. The electrons are accelerated using electrical charge, toward a series of electrodes (dynodes) arranged in a zigzag pattern. The dynode is made from a material that ejects more electrons than hit it at any given moment. By the time the electrons are ejected by the last dynode, there are a whole lot more electrons than started out. This is how the signal is amplified. Finally, the electrons hit the anode on the far end of the PMT and register an electrical signal. The energy of the original signal is preserved through the PMT, but some precision is lost. Therefore, a drawback to using a PMT is that energy resolution is poor. This means that if energy vs. count data are plotted (a spectrum!), specific photon energies will be seen as broad peaks.

6.2.2 INORGANIC SCINTILLATORS

Inorganic scintillators are salts (inorganic compounds) that light up when hit by ionizing radiation. The most commonly used are crystals of sodium iodide (NaI) or cesium iodide (CsI). In both cases, the crystals light up a lot better if they are doped with a tiny amount of thallium (Tl). The doped crystalline material is represented as NaI(Tl) or CsI(Tl).

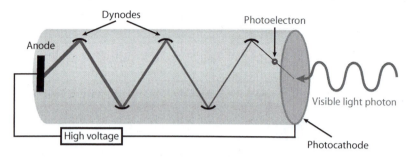

FIGURE 6.5   Schematic of a photomultiplier tube (PMT).

Notice that there is a lot of iodide in both detectors. Iodine is a relatively high Z element which means that it has a higher probability of interacting with photons, especially via the photoelectric effect. These materials are also very dense when compared to the gases used in gas-filled detectors. As a result, inorganic scintillators are much more efficient at detecting X- and γ-rays.

Figure 6.6 demonstrates how a typical NaI(Tl) detector works. The crystal is cylindrically shaped and sealed inside a shiny metal case. The case is reasonably thin, so that X- and γ-rays can easily penetrate it. Once inside the crystal, the photon interacts, most likely with the iodide inside the crystal, eventually creating visible (scintillation) light. The nature of the interaction is discussed in more detail in section 6.4. The light then goes directly toward the PMT, or it is reflected off the shiny metal enclosure into the PMT.

After the PMT amplifies the signal, it is run through a multichannel analyzer (MCA), which sorts the signal by energy (channel, each channel covers a discrete energy range). Despite some loss of energy resolution by the PMT, both count and energy data are obtained.

6.2.3 ORGANIC SCINTILLATORS

All of the detection methods mentioned so far suffer from the fact that they can only detect radiation that enters the detector (and interacts). Since radiation is typically emitted in all directions equally, less than 50% of all emitted radiation can be expected to enter the detectors. If we have a low activity sample, especially if it emits low energy radiation, it would be better if we could place the sample inside our detector, so that we could detect all decays. This can be done with an **organic scintillator**, as shown in figure 6.7.

Organic scintillation works by dissolving the radioactive material in a "scintillation cocktail." This cocktail is made up of organic (low Z) materials: a solvent, a scintillator, and sometimes other stuff to help make it glow brighter or with the right wavelength. Low Z materials are used to minimize bremsstrahlung that can occur when beta particles are emitted during decay. This solution is placed in a glass or plastic vial, which is put in a dark place (inside an instrument!) with PMTs to watch it and record data.

First the decay takes place. Since the cocktail is mostly solvent, it is most likely that the solvent absorbs energy from the ionizing radiation. The solvent then transfers this energy to a scintillator molecule, which becomes excited. The scintillator then

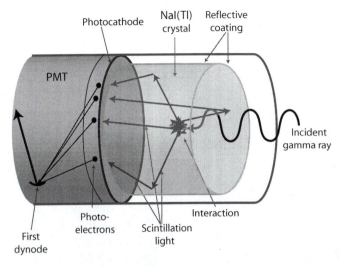

FIGURE 6.6   A NaI(Tl) detector at work.

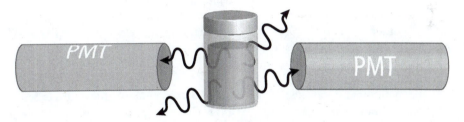

FIGURE 6.7   Organic scintillation gives off photons of visible light.

de-excites, emitting a photon. The photon hits a PMT, which amplifies the signal and records it.

. As mentioned above, this is a great way to measure samples emitting low-energy radiation, such as weak beta-emitting nuclides. It is not very efficient for hard X-rays or γ-rays, as these photons are unlikely to interact with the low Z materials inside the vial.

Another drawback is **quenching**. Quenching here is a different issue than for gas-filled detectors. In terms of scintillation counting, it refers to anything that reduces light output. It can be a fingerprint or other smudge on the outside of the vial. A highly colored sample (like blood) could also be a problem. If the sample volume is significant compared to the volume of the cocktail, the photon output will be reduced. Finally, O_2 can inhibit the transfer of energy from the solvent to the scintillator, and must be purged from the solution.

6.3   OTHER DETECTORS

6.3.1   SEMICONDUCTOR DETECTORS

Semiconductor detectors are very similar to gas-filled detectors, except they are commonly filled with a silicon (Si) or germanium (Ge) crystal instead of a gas.

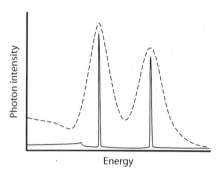

FIGURE 6.8   Comparison of the gamma spectrum energy resolution of inorganic scintilla-tion (dashed line) and semiconductor (solid line) detectors.

Therefore, the detector is ~2000–5000 times as dense and much more efficient at detecting photons. Like the inorganic scintillators, these crystals work better when doped with a small amount of an impurity—in this case lithium (Li). These materials are then represented as Si(Li) ("silly") and Ge(Li) ("jelly").

Si and Ge are metalloids on the periodic table. This means they have chemical and physical properties that are intermediate between the metals and the non-metals. Most important for us is the fact that they are neither conductors of electricity, nor are they insulators—they are semiconductors. This means that sandwiching a crystal between two electrodes will allow the ion pairs produced from ionizing radiation to migrate to the electrodes.

The main disadvantages of semiconductor detectors are that they are expensive to make and need to be cooled to ~100 K (liquid nitrogen!) in order to work. The main advantage is that their energy resolution is excellent, as illustrated in figure 6.8.

6.3.2   THERMOLUMINESCENT DOSIMETERS

Certain salts (inorganic compounds) can be used to store information on exposure to ionizing radiation. Good materials for this are lithium fluoride, LiF, and calcium fluoride doped with a little manganese, $CaF_2(Mn)$. When electrons in crystals of these salts are excited by interaction with ionizing radiation, they get stuck (trapped) in intermediate energy levels. They can stay trapped for months or even years! When the crystal is heated up, the electrons are freed from their traps, and return to the ground state. Along the way they emit visible light, which can be recorded by a PMT. The intensity of the light emitted is proportional to the amount of radiation the crystal was exposed to. This process (**thermoluminescence**) is illustrated in figure 6.9.

These crystals are typically built into badges worn by personnel working with radioactive material or radiation-producing devices to estimate their occupational dose. These badges are called **thermoluminescent dosimeters** (TLDs), and can be collected on a monthly or quarterly basis to ensure workers are not exposed to too much radiation. Typically three or four crystals will be placed inside a single badge, with different levels of shielding. This way the penetrating ability of the radiation the worker is exposed to can be estimated.

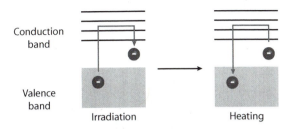

FIGURE 6.9   Thermoluminescence.

6.4   GAMMA SPECTROSCOPY

Gamma spectroscopy is the ability to use detectors to collect both count and energy information from γ-rays. By plotting energy vs. counts, we can see different energy γ-rays at the same time. It is sometimes referred to as Pulse-Height or Energy Spectrometry. Regardless of the name, it is a wonderful tool for discerning various radioactive nuclides. Unfortunately the information is sometimes complicated by what appear to be odd features in the spectrum. Fortunately these features can all be understood in terms of the interactions of high-energy photons with matter discussed in Chapter 5.

As you already know, γ-rays are very penetrating, that is, they have a low probability of interacting with matter. We can boost this probability by making them try to pass through relatively dense matter. That's why lead (Pb) is commonly used for shielding high-energy photons. If we want to detect them, we'll also have to use fairly high Z material. As discussed above, that's why inorganic scintillators or semiconductor detectors are commonly used. Remember that these detection methods give count *and* energy information from gamma emitting sources! Instead of just getting count data, we get count data spread out over the gamma energy spectrum, like figure 6.10 for ^{137}Cs.

But wait a minute! We know that the decay diagram for ^{137}Cs (fig. 6.11) only shows one γ-ray being emitted, with an energy of 662 keV. Since nuclear energy states are quantized, we should *only* see counts for 662 keV. Why is the peak at 662 keV so broad, and what's all the other stuff in the spectrum? Why is there a broad continuum of counts below ~500 keV, what is the bump around 200 keV, and the really big peak at relatively low energy? The answers to these important questions (and maybe some others!) have to do with how γ-rays interact with matter, because that's how they are detected.

Figure 5.9 tells us that if we build a detector out of low or medium Z matter, most of the gamma photons that interact with it will do so via the Compton effect. If we make it with high Z materials, the photoelectric effect will dominate, at least for low-energy γ-rays.

Does it make a difference whether the photon interacts with the iodide in the detector via the photoelectric or Compton effect? You betcha. When a photon interacts via the photoelectric effect, all of its energy is dumped into the detector, giving us a signal for the energy of that photon. In the γ-ray spectrum this is called the **photopeak** (or the total energy peak, or the gamma peak). This is what we want to

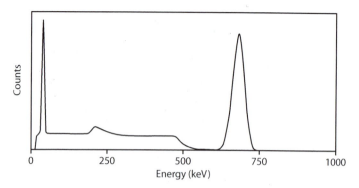

FIGURE 6.10   Gamma spectrum for ^{137}Cs using a NaI(Tl) detector.

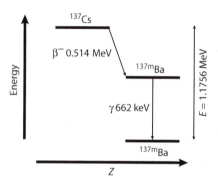

FIGURE 6.11   Decay diagram for ^{137}Cs.

happen; therefore we make detectors out of high Z materials. As noted in Chapter 5, high Z also means a higher overall probability of interaction, meaning fewer gamma photons will pass completely through the detector without interacting.

If the photoelectric effect gives the energy of the gamma photon, why is the photopeak in figure 6.10 so broad? As explained earlier in this chapter, it is because a PMT is used to amplify the signal. During this process, the energy information gets blurred, resulting in a broad peak (poor energy resolution) on the spectrum.

When the gamma photon interacts with the detector via the Compton effect, the photon *could* deposit only some of its energy in the detector then leave, taking with it some of its original energy as the scattered photon. Since the amount of energy transferred by the photon undergoing Compton scattering is variable, then different amounts of energy will be deposited in the detector for each photon undergoing the Compton effect then escaping the detector. This will give rise to a broad continuum of counts called the **Compton continuum**. This is the broad plateau of counts in figure 6.10. One way to decrease the relative number of counts under the Compton continuum relative to the photo peak is to increase the size of the detector. A larger detector will give the scattered photon more opportunities to interact, leaving all of the incident photon's energy inside the detector.

What's the maximum energy Compton scattering can leave in the detector? Assuming the gamma photon only undergoes one Compton scattering interaction inside the detector, it would be when it is scattered through 180°. This energy can be calculated based on the original energy of the photon using equation (6.1):

$$E_{CE} = \frac{E_0^2}{E_0 + 0.2555} \tag{6.1}$$

where E_0 is the energy of the incident photon and E_{CE} is the "maximum energy" that can be deposited in the detector. Both energy values must have units of MeV for this equation to be true. E_{CE} is also known as the **Compton edge**. The Compton edge for the 662 keV photopeak observed in the ^{137}Cs spectrum is:

$$E_{CE} = \frac{0.662^2}{0.662 + 0.2555} = 0.478 \text{ MeV} = 478 \text{ keV}$$

Careful examination of figure 6.10 shows that the Compton edge does not define a strict upper limit to the continuum, but rather it is the point where counts begin to decrease with increasing energy. Why are some Compton continuum counts still observed above the Compton edge? This is partly due to the poor energy resolution of NaI(Tl) detectors. It can also result from a gamma photon having more than one Compton scattering interaction inside the detector and exiting leaving slightly more energy than is calculated for E_{CE}.

Compton scattering can also take place outside the detector. A gamma photon originally on its way away from the detector could be scattered off an air molecule, you, your lab partner, or some other matter. It shows up in the gamma spectrum with a specific energy. The energy has a certain value because the scattering angle has to be 180° (or pretty close to it) for the scattered photon to end up in the detector. The energy of this small peak (E_{BS}, aka the **backscatter** peak) is given by the following, where all energy values are again given in MeV:

$$E_{BS} = \frac{E_0}{1 + (3.91 \times E_0)} \tag{6.2}$$

Despite the fact that they originate from different phenomena, there's a simple mathematical relationship between E_{BS} and E_{CE}.

$$E_0 = E_{CE} + E_{BS} \tag{6.3}$$

This equation is really the same as equation (5.9):

$$E_0 = E_{SC} + E_{CE}$$

where E_{CE} is the energy of the Compton electron.

^{137}Cs emits a 0.662 MeV γ-ray, therefore its backscatter peak should be found at:

$$E_{BS} = \frac{0.662}{1 + 3.91(0.662)} = 0.184 \text{ MeV} = 184 \text{ keV}$$

We now understand most of what we see in the γ spectrum for ^{137}Cs. These features are labeled in figure 6.12.

So what's up with the low energy peak, and why is it labeled Ba X-ray? The ^{137}Ba isomer (a metastable state) that is formed after ^{137}Cs emits a beta particle (see fig. 6.11) can also decay to the ground state by spitting out a conversion electron. The conversion coefficient is $\alpha_K = 0.093$, which means that ∼1/10 of the time a conversion electron is ejected. The resulting hole in the K-shell of ^{137}Ba can be filled by an electron from an L shell, which could generate an X-ray with an energy of 32 keV. These X-ray photons are also detected by the NaI crystal, producing the low energy peak observed in the gamma spectrum of ^{137}Cs.

Equipped with a solid understanding of the interactions of γ-rays with matter (Chapter 5), let's see if we can understand what other odd features we might come across in gamma spectra.

The relative intensities of gamma peaks can depend on a few factors. As we already know, higher-energy photons have a lower probability of interacting with matter. Therefore, we would expect to see smaller peaks for higher-energy photons even if the source is emitting equal numbers of low- and high-energy photons. Intensities of two different peaks may also vary if they are part of two different decay branches or if they have significantly different conversion coefficients. Finally, relative intensities can vary due to the relative probability that each photon will be emitted as part of the decay.

A great example is ^{57}Co. As illustrated in figure 6.13, it decays via electron capture to an excited state of ^{57}Fe with a 5/2 spin state. This excited state has two options. It can decay directly to the ground state, emitting a 136 keV gamma

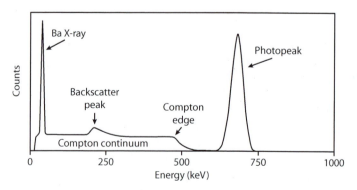

FIGURE 6.12   Labeled gamma spectrum for ^{137}Cs.

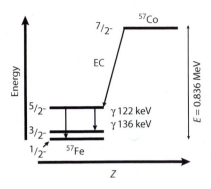

FIGURE 6.13 Decay diagram for ^{57}Co.

photon, or it can decay to a lower-energy excited state emitting a 122 keV gamma photon. The 122 keV gamma photon is more likely to be observed (more intense in spectrum) than the 136 keV gamma photon because the change in nuclear spin states is less dramatic. We would also expect the 136 keV peak to be smaller because it is higher in energy than the 122 keV peak. The relative intensities are observed to be 85.6:10.7.

With low-energy gamma photons, it is *sometimes* possible to observe a small peak that is 28 keV lower in energy than the photopeak. An example spectrum is shown in figure 6.14.

A low-energy gamma photon will most likely interact with iodide (I^-) in the detector via the photoelectric effect. This creates an electron vacancy in the K shell of the iodide. When an electron from a higher-energy shell fills this vacancy, X-rays are generated. Normally the crystal absorbs the energy of this X-ray, that is, it interacts before it can get out of the crystal. If a number of these find their way out of the detector, the recorded energy will be missing 28 keV (the energy of the I K_α X-ray). This is called an **iodine escape peak**.

It is only observed with low-energy gamma photons, since they are most likely to interact with I^- via the photoelectric effect. Low-energy photons also have a higher probability of interacting with matter, in general. Therefore, odds are better that they will interact close to the surface of the detector. This increases the probability that the K_α X-ray will escape the detector, since it (potentially) has less matter to travel through to get out.

Figure 6.15 shows the decay diagram for ^{60}Co. After beta emission, an excited state of ^{60}Ni is formed. Like the ^{57}Co example above, this excited state has two choices; decay to another excited state or go all the way to the ground state. In this case the change in spin state is too dramatic to allow direct decay to the ground state. As a result, photopeaks are observed at 1.17 and 1.33 MeV. If the sample is reasonably toasty (high activity), a small peak may be observed at 2.50 MeV. This is not due to a $4+ \rightarrow 0+$ transition, but is a **sum peak**. This happens when two different γ photons hit the detector at the same time, and are therefore recorded as a single count.

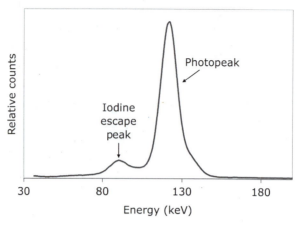

FIGURE 6.14 Gamma spectrum of ^{57}Co.

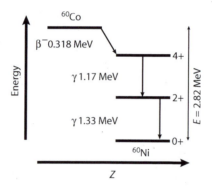

FIGURE 6.15 Decay diagram for ^{60}Co.

Sum peaks become more probable when a well detector is used. A cutaway diagram of a well detector is shown in figure 6.16. Basically, it's a regular detector with a hole drilled into it so the source can be placed in the center of the crystal.

Figure 6.17 shows the decay diagram for ^{22}Na. As expected, its gamma spectrum has a photopeak at 1.275 MeV. A large peak is also observed at 0.511 MeV. This peak is due to β^+ annihilation (with an electron) and is called an **annihilation peak**. The positron most likely annihilates outside of the detector, generating two 511 keV photons. A number of these photons enter the detector and are recorded at 0.511 MeV. The ^{22}Na spectrum can also exhibit sum peaks at 1.02 MeV (two annihilation photons entering the detector at the same time), and at 1.786 MeV (1.275 + 0.511 MeV).

If the energy of the gamma photon is >1.022 MeV (as in the last two examples), a peak at 0.511 MeV may be observed as a result of pair production *outside* the detector crystal. The positron produced annihilates outside the detector and sends

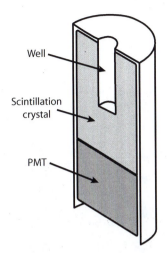

FIGURE 6.16   Cutaway view of a well detector.

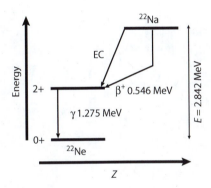

FIGURE 6.17   Decay diagram for ^{22}Na.

one of the annihilation photons into the detector. This requires a relatively high activity source.

What if pair production takes place inside the detector? Then the annihilation photons are created inside the detector. If one or both get out of the detector without interacting, then counts will be recorded with an energy of 0.511 or 1.022 MeV below the photopeak. The peak that is 511 keV below the photopeak is called an **escape peak**. The peak 1.022 MeV below the photopeak is called a **double escape peak**. An example of a spectrum exhibiting escape peaks is shown in figure 6.18. For these peaks to be observed, the photopeak must have an energy well above the 1.022 MeV threshold for pair production.

If lead is used for shielding or as a collimator, a peak at ~80–90 keV could be observed. If a gamma photon from the radioactive source interacts with Pb via the photoelectric effect, it can create a vacancy in the Pb K electron shell. When this

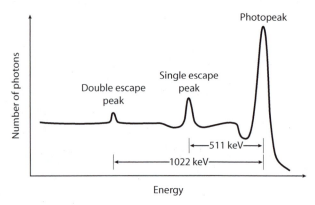

FIGURE 6.18   A gamma spectrum with escape peaks.

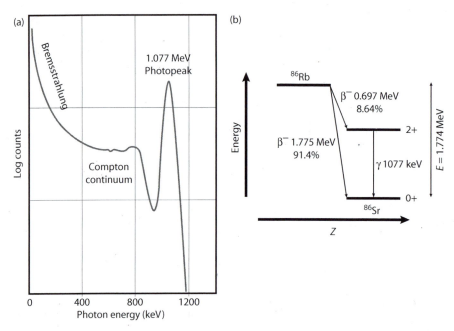

FIGURE 6.19   (a) Gamma spectrum (adapted from Heath, R.L. *Scintillation Spectrometry: Gamma-Ray Spectrum Catalogue,* 2nd ed., Idaho National Laboratory, Idaho Falls, 1997, http://www.inl.gov/gammaray/catalogs/catalogs.shtml.) and (b) decay diagram for ^{86}Rb.

vacancy is filled with an electron from a higher energy shell, a Pb characteristic X-ray can be produced. The X-ray can then enter the detector and record a count for its energy. This results in a small peak at ∼75 keV. This peak is called the **Pb X-ray** peak.

When *a lot* of beta particles, relative to the number of gamma photons, are emitted, **bremsstrahlung** is observed. Remember that bremsstrahlung is the production of X-rays as a high-speed electron slows down while interacting with matter. These

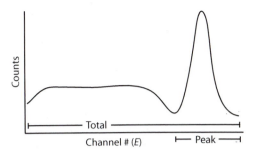

FIGURE 6.20   Peak-to-total ratio.

interactions are highly variable, and can produce a broad energy spectrum of X-ray photons. Bremsstrahlung can occur either inside or outside the detector. All that matters to us is whether the detector records the X-ray. If it does, it'll show up as a count at the low-energy end of the gamma spectrum. Collectively, it is manifested in the far left of the gamma spectrum as a slow rise in the count rate as photon energy decreases. Figure 6.19 gives an example spectrum and the corresponding decay diagram.

The decay diagram for ^{86}Rb (fig. 6.19b) shows that the beta decay branch ratio leading to gamma emission is small. Mostly, this nuclide is spitting out beta particles. As a result, some bremsstrahlung is also observed in its gamma spectrum (fig. 6.19a).

If something is between the source and the detector, the gamma photons *can* undergo Compton scattering while traveling through the object on their way into the detector. This is known (logically!) as **object scattering**. Since Compton scattering lowers the energy of the photon, it will be recorded under the Compton continuum. Object scattering leads to more counts observed under the Compton continuum and fewer counts under the photopeak.

Peak-to-total ratio is a measure of detector quality. It is determined by summing the counts for all the channels under the total-energy peak and dividing by the sum of all the counts for the channels under the Compton continuum and those for the peak (fig. 6.20). Peak-to-total ratio will depend on the energy of the γ-ray. The ratio generally decreases with increasing energy of the γ-ray. This makes sense as the *relative* probability of Compton scattering increases with increasing γ-ray energy. Overall, the probability of interaction goes down as gamma energy increases, but the odds of undergoing the photoelectric effect go down a lot faster than the odds of Compton scattering (fig 5.9).

QUESTIONS

1. Using the information from Question 2 in Chapter 5, determine what sorts of electronic transitions in argon atoms lead to the production of X-rays, UV, and visible light.
2. Would a GM counter be efficient at detecting neutrinos? Briefly explain.

3. What would be the best detector to obtain count and energy information from a beta emitter? A gamma emitter? Assume only one energy particle or photon is emitted in each case.

4. A nuclear medicine technologist working in a radiopharmacy performs swipe tests weekly to look for areas of radioactive contamination. The swipe test is performed by wiping a small piece of paper (like filter paper) across the surface to be tested, placing it in a glass test tube, then placing the assembly into a gamma spectrometer. If the only radionuclide in use at this pharmacy is ^{99m}Tc, comment on why this might not be an especially good quality control (QC) procedure.

5. Radiation therapists work with equipment that generates high-energy photons (X-rays). If they wanted to determine the monthly dose from small amounts of radiation "leaking" from their machines (*not* the main beam) into the treatment room, what would be the best way to go about it?

6. A 0.100 MeV γ-ray interacts inside a NaI(Tl) detector by the photoelectric effect. What element is most likely involved in this reaction? What is the atomic shell most likely to lose an electron from this element? Briefly explain how an iodine escape peak might be formed from this γ-ray. Calculate the energy of the freed electron.

7. Calculate the energy of the backscatter peak and the Compton edge for the γ-ray most commonly observed in the decay for ^{99m}Tc. Sketch the γ-ray spectrum that would be observed for this γ-ray, labeling the important features.

8. During a routine survey of the crab nebula, the science staff of the Stargarzer discovers an abundance of one element. Sensor analysis shows it to be Al, P, Cu, Br, or Tm. You obtain a small sample and place it in a neutron beam generated by one of the EPS conduits. Upon removal, you detect a 0.51 MeV γ-ray using a flat-faced NaI(Tl) detector. The same sample shows gamma peaks at 0.51 and 1.02 MeV when placed in a well detector. Using either detector, at least one other peak is observed in the gamma spectrum, but interference from the nebula is preventing you from observing their energies. Your report on this element is due to the ship's science officer in one hour. What can you contribute to the solution of this mystery? What experiment would you suggest be done next?

9. Explain the origin of the Compton continuum often observed in gamma spectra.

10. When a lead collimator is used with a gamma camera, a peak at 75 keV is observed. When the collimator is absent, no peak is seen at 75 keV. Briefly explain the origin of this peak.

11. Why are Compton scattering and pair production undesirable interactions inside of a gamma detector? How are detectors designed to minimize these interactions?

12. The dashed line in the figure shown here (fig. 6.21) represents that gamma spectrum of a radionuclide in air, while the solid line is the spectrum for the same source inside of a patient. Briefly explain the differences between the two spectra. What happens to the peak-to-total ratio as this source is put inside of a patient?

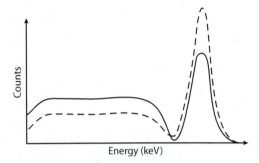

FIGURE 6.21   Gamma spectra outside (dashed line) and inside (solid line) a patient.

7 Nuclear Reactions

Radioactive decay is not considered a nuclear reaction. Some readers may consider this strange because of the obvious analogy to chemical reactions. One compound breaking up to form two or more compounds is called a decomposition reaction by chemists, and considered no less a chemical reaction than an acid–base neutralization reaction, or any other type of chemical reaction. So why is nuclear decay not considered a reaction?

Nuclear scientists define nuclear reactions as the collision of two nuclei, or a nucleus and a neutron, resulting in the production of a different nuclide. This definition does not include decay processes, which could be defined as the spontaneous transformation of one nuclide into another, usually accompanied by the nuclear emission of a particle or photon. This may all seem like a relatively pointless semantic debate, but the reader should be aware that it is generally considered improper to refer to decay as a reaction.

The nuclear reactions we'll study in this chapter can be generically represented by the following:

$$A + x \rightarrow B + y$$

where A is the target nuclide, x is the **projectile**, B is the **product** nuclide, and y is an emitted particle or photon. The projectile x and emitted y can be a gamma photon or a wide range of particles, such as an alpha particle (α or ^{4}He), neutron (n), proton (p), deuterium (d or ^{2}H), or tritium (t or ^{3}H). Curiously enough, something is always spit out; the target and projectile do not simply combine to form a new nuclide. We can, however, imagine the temporary formation of C from the addition of x to A. Inserting C into our generic reaction above gives:

$$A + x \rightarrow [C]^* \rightarrow B + y$$

The square brackets around C indicate its transient nature, meaning that it doesn't exist for reasonable amounts of time. The asterisk indicates that this nuclide is in an excited state, typically much higher in energy than the ground state. Taken together, [C]* represents what is called a **compound nucleus**.

The following shorthand can represent the above nuclear reaction:

$$A(x, y)B$$

where the target and product nuclides are placed outside the parentheses, and the projectile and emission are placed inside. As a result, these reactions are often referred to as "**x,y-type reactions**." From this generic point of view this may not seem like much of a shorthand. Hopefully, its utility will be illustrated in the examples below.

Neutron capture is a fairly common nuclear reaction. An example and its short-hand are given below.

$$^{51}_{23}V_{28} + ^{1}_{0}n \rightarrow ^{52}_{23}V_{29} + \gamma$$

$$^{51}V(n,\gamma)^{52}V$$

7.1 ENERGETICS

Since a gamma photon is emitted in the reaction above, this is an example of an n-gamma reaction. Notice that numbers of nucleons are conserved in this, as with all nuclear transformations. Are mass and energy collectively conserved? We can find out by looking at the difference in mass between reactants and products, just like we did with decay. Let's do the math!

Mass of reactants	^{51}V	50.943964 u
	^{1}n	$+1.008665$ u
		51.952629 u
Mass of product	^{52}V	-51.944779 u
Reactants−products	^{51}V	0.007850 u

Note that atomic masses are used for ^{51}V and ^{52}V, just as we saw with decay, and mass is lost in this n-γ reaction, therefore energy is produced. The variable Q is used to designate energy in nuclear reactions, instead of the E that is used for decay. While potentially confusing, it serves to further differentiate these two types of nuclear processes. Let's calculate how much energy is produced by this reaction:

$$Q = \frac{931.5 \, \text{MeV}}{u} \times 0.007850 \, u = 7.312 \, \text{MeV}$$

When mass is lost in the course of the reaction, then $Q > 0$ and the reaction is termed **exoergic** (aka exothermic). Decay is always exoergic, but that won't be true for nuclear reactions. Occasionally, we will run into reactions where mass is gained, Q is negative, and the reaction is termed **endoergic** (or endothermic).

Dealing with energy in nuclear reactions is much more complex than with decay processes. For example, in the n-γ reaction above, the energy produced comes from the binding of an additional neutron to a ^{51}V nucleus and is released through the gamma photon and kinetic energy of the product nuclide. There's also energy on the reactant side that we need to be concerned with—the kinetic energy of the projectile is also nonzero. In some reactions, the projectile *must* have a certain amount of energy before a reaction will proceed. How can we sort all of this out? We'll get to all of this eventually; first, let's consider the projectile.

The projectile in our example above is a neutron. Neutrons make great projectiles because they have no electrical charge. As a result, they are neither repelled from nor attracted to an atom's protons or electrons. This means a neutron can approach a nucleus without hindrance. Positively charged projectiles do not have the same advantage. As they approach a nucleus they will begin to feel Coulomb repulsion of the positively charged nucleus.

The first example of a charged particle being used in a nuclear reaction was discovered by Ernest Rutherford in 1919!

$$^{14}_{7}N + ^{4}_{2}He \rightarrow ^{17}_{8}O + ^{1}_{1}H$$

In this reaction ^{14}N is the target, an alpha particle is the projectile, ^{17}O is the product, and a proton is emitted. This is an "alpha-p" reaction and can also be represented as $^{14}N(\alpha, p)^{17}O$. Let's take a detailed look at the energetics of this reaction, starting by calculating the energy released (or required!).

Mass of reactants	^{14}N	14.003074 u
	^{4}He	+4.002603 u
		18.005677 u
Mass of products	^{17}O	16.999132 u
	^{1}H	+1.007825 u
		18.006957 u

Reactants−products $= 18.005677 - 18.006957 = -0.001280$ u

Mass is gained in this reaction, so energy needs to be put in—this is an endoergic reaction. Nowadays this is usually accomplished by accelerating the projectile using sophisticated equipment. Rutherford was able to accomplish this in 1919 by using an alpha particle projectile that was already packing a punch. The alpha particles were generated from the decay of ^{214}Po and have an energy of 7.68 MeV. How much energy does this reaction require?

$$Q = \frac{931.5 \text{ MeV}}{u} \times \left(-0.001280 \text{ u}\right) = -1.19 \text{ MeV}$$

The Q value is the minimum energy required by this reaction to manufacture the necessary mass. We also need to be concerned with the energy, or momentum, of the products. Both mass and momentum are accounted for in the **threshold energy** (E_{tr}):

$$E_{tr} = -Q \times \left(\frac{A_A + A_X}{A_A}\right) \quad (7.1)$$

where A_A is the mass number of the target, A_X is the mass number of the projectile, and A_B is the mass number of the product. Note the negative sign in front of Q. This equation only makes sense when Q is negative, that is, the reaction is endoergic. For Rutherford's reaction:

$$E_{tr} = 1.19 \text{ MeV} \times \left(\frac{14+4}{14}\right) = 1.53 \text{ MeV}$$

This tells us the alpha particle needs a minimum of 1.53 MeV of energy for this reaction to produce the required mass and momentum of the products. While this is greater than the 1.19 MeV required solely for matter creation, it is still quite a bit less than the 7.68 MeV of the ^{214}Po alpha particle. Any excess energy brought in by the projectile will end up as kinetic or excitation energy of the products. In other words, the products are movin' or shakin' more than they would otherwise.

Great! We've got Rutherford's reaction under control, right? Not quite. We need to be concerned about the fact that we're bringing together two positively charged particles. They will try to repel each other (Coulomb repulsion) with increasing force as they get closer and closer together. The projectile must have a certain amount of energy to overcome this repulsion. The minimum amount of energy required to overcome this **Coulomb barrier** (E_{cb}) can be approximated as:

$$E_{cb} \approx 1.11 \times \left(\frac{Z_A Z_X}{A_A^{1/3} + A_X^{1/3}} \right) \text{ (in MeV)} \tag{7.2}$$

where Z_A and Z_X are the atomic numbers for the target and projectile, respectively. This energy doesn't get soaked up and doesn't enter into our energy accounting for the reaction; it is simply a barrier to overcome. Just as we did with Q, we have to also be concerned with conservation of momentum here. The minimum energy required to overcome the **effective Coulomb barrier** and provide enough energy for the momentum of the products is E_{ecb} (in MeV).

$$E_{ecb} \approx 1.11 \times \left(\frac{A_A + A_X}{A_A} \right) \times \left(\frac{Z_A Z_X}{A_A^{1/3} + A_X^{1/3}} \right) \tag{7.3}$$

For Rutherford's reaction, the effective Coulomb barrier is:

$$E_{ecb} \approx 1.11 \times \left(\frac{14 + 4}{14} \right) \times \left(\frac{7 \times 2}{\sqrt[3]{14} + \sqrt[3]{4}} \right) = 4.95 \text{ MeV}$$

For endoergic reactions, the minimum amount of energy required to make it proceed in good yield is the greater of the threshold energy and the effective Coulomb barrier, with the effective Coulomb barrier usually being the larger of the two. The alpha particles in this example have ample energy (7.68 MeV) to make it over this hump.

Finally, we should also be concerned about aim. Projectiles don't always hit their targets dead-on. Glancing blows can sometimes be productive, but problems with conserving angular momentum come into play. This concern is beyond the scope of this text.

Let's take a step back and look at this reaction as two separate steps. The first step is the formation of the compound nucleus ($[^{18}F]^*$), followed by its decomposition into the ^{17}O product and a proton.

$$^{14}_{7}N + {}^{4}_{2}He \rightarrow \left[{}^{18}_{9}F \right]^* \rightarrow {}^{17}_{8}O + {}^{1}_{1}H$$

The kinetic energy (*not* excitation energy) of the compound nucleus is approximated by:

$$E_{KC} \approx E_{KX} \times \left(A_X / A_C \right) \tag{7.4}$$

where E_{KX} is the kinetic energy of the projectile. Note that this equation applies to *any* nuclear reaction (regardless of the sign of Q). Applying this equation to Rutherford's reaction:

$$E_{KC} \approx 7.68 \text{ MeV} \times (4/18) = 1.71 \text{ MeV}$$

we can use this information to calculate just how far $[^{18}F]^*$ is above ground state ^{18}F. Consider the (theoretical) reaction:

$$^{14}N + {^4}He \rightarrow {^{18}}F$$

Let's also assume (in theory!) that all of the above are in the ground state. How much energy is produced or consumed? In other words, what is Q?

$$Q = (14.003074 \text{ u} + 4.002603 \text{ u} - 18.000938 \text{ u}) \times \frac{931.5 \text{ MeV}}{u}$$

$$= +4.41 \text{ MeV}$$

Therefore, when the Rutherford reaction takes its first step and forms a compound nucleus, 4.41 MeV of energy is produced. Remember that the alpha particle already has 7.68 MeV of energy. The excitation energy of the $[^{18}F]^*$ in the Rutherford reaction is therefore the energy of the projectile, plus the energy released for the first step of the reaction minus the kinetic energy of the compound nucleus:

$$7.68 \text{ MeV} + 4.41 \text{ MeV} - 1.71 \text{ MeV} = 10.38 \text{ MeV}$$

That's one excited compound nucleus! It's no wonder it can't even hold itself together. If you're clever enough, you can now calculate the energy (excitation plus kinetic) of the products, but this question can be answered without the hassle of calculating the excitation energy of the compound nucleus:

$$7.68 \text{ MeV} - 1.19 \text{ MeV} = 6.49 \text{ MeV}$$

Where the 7.68 MeV is still the energy that the alpha brings to the reaction, the 1.19 MeV is the energy needed to make the mass (Q, from a couple pages back), and 6.49 MeV is the energy (excitation plus kinetic) of the products.

With all this information, we can draw a tidy little diagram that shows how energy changes during the course of this reaction (fig. 7.1). We know that going from ground state reactants to ground state products costs 1.19 MeV of energy (Q), so the

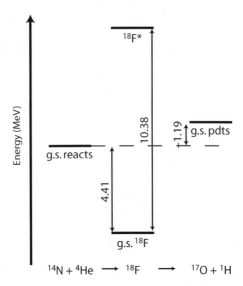

FIGURE 7.1 Energy diagram for Rutherford's reaction.

products should be slightly higher in energy than the reactants. Note that the products of nuclear reactions are rarely generated in their ground state. They typically have some level of excitation and kinetic energy. We can also show the ground state, and actual state of the compound nucleus, since 4.41 MeV is released in the formation of the compound nucleus and it is formed in an excited state that is 10.38 MeV above that ground state.

   Example: What is the minimum energy required to make the following reaction proceed in good yield?

$$^{240}_{94}\text{Pu} + ^{4}_{2}\text{He} \rightarrow ^{243}_{96}\text{Cm} + ^{1}\text{n}$$

This is an α-n reaction—$^{240}\text{Pu}(\alpha,n)^{243}\text{Cm}$. First find the difference in mass between reactants and products:

Mass of reactants	^{240}Pu	240.053807 u
	^{4}He	+4.002603 u
		244.056410 u
Mass of products	^{243}Cm	243.061382 u
	^{1}n	+1.008665 u
		244.070047 u

Reactants − products = 244.056410 − 244.070047 = −0.013637 u

Once again, mass is gained, therefore this is an endoergic reaction. The energy required to make the mass needed for this reaction (Q) is:

$$Q = \frac{931.5 \text{ MeV}}{\text{u}} \times \left(-0.013637 \text{ u}\right) = -12.7 \text{ MeV}$$

Since it is endoergic, we need to calculate threshold energy (mass + momentum):

$$E_{tr} = 12.7 \text{ MeV} \times \left(\frac{240+4}{240}\right) = 12.9 \text{ MeV}$$

Since it involves the collision of two nuclei, we also should calculate the effective Coulomb barrier:

$$E_{ecb} \approx 1.11 \times \left(\frac{240+4}{240}\right) \times \left(\frac{94 \times 2}{\sqrt[3]{240} + \sqrt[3]{4}}\right) = 27.2 \text{ MeV}$$

This is a lot higher than the threshold energy and is therefore the minimum amount of energy required to make this reaction proceed in good yield. Alpha particles will need to be moving pretty fast to make this reaction work.

In both of the examples we've looked at so far, the effective Coulomb barrier was higher than the threshold energy and, therefore, determined the minimum energy required for the reaction. The effective Coulomb barrier will vary somewhat consistently with the masses of the colliding objects as illustrated in figure 7.2. The two lines show how the effective Coulomb barrier changes for ^{1}H and ^{4}He as projectile while the atomic number of the target nucleus increases. Only data for the most abundant, stable, target nuclides were used to generate this figure.

The bigger the projectile and target are, the harder it is to get them together. This makes sense as the amount of positive charge generally increases with size. This is nicely illustrated in our two examples. In Rutherford's reaction we slammed an

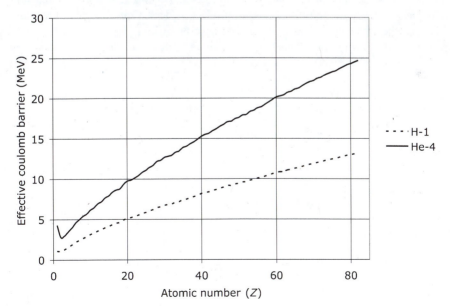

FIGURE 7.2   Variations in the effective Coulomb barrier with projectile and target nucleus.

alpha particle (two protons) into a ^{14}N (7 protons) atom and saw an effective Coulomb barrier of ~5 MeV. In the second example, the effective Coulomb barrier jumped to ~27 MeV because we ran an alpha particle into a ^{240}Pu (94 protons!) nucleus.

Let's approach nuclear reactions from another perspective. Let's say we wanted to make some ^{18}F (commonly used in PET imaging). What possible reactions could be used?

1   $^{15}_{7}\text{N} + ^{4}_{2}\text{He} \rightarrow ^{18}_{9}\text{F} + ^{1}\text{n}$ $^{15}\text{N}(\alpha,\text{n})^{18}\text{F}$

2   $^{16}_{8}\text{O} + ^{3}_{1}\text{H} \rightarrow ^{18}_{9}\text{F} + ^{1}\text{n}$ $^{16}\text{O}(\text{t},\text{n})^{18}\text{F}$

3   $^{17}_{8}\text{O} + ^{2}_{1}\text{H} \rightarrow ^{18}_{9}\text{F} + ^{1}\text{n}$ $^{17}\text{O}(\text{d},\text{n})^{18}\text{F}$

4   $^{18}_{8}\text{O} + ^{1}_{1}\text{H} \rightarrow ^{18}_{9}\text{F} + ^{1}\text{n}$ $^{18}\text{O}(\text{p},\text{n})^{18}\text{F}$

5   $^{19}_{9}\text{F} + ^{1}_{1}\text{H} \rightarrow ^{18}_{9}\text{F} + ^{1}_{1}\text{H} + ^{1}\text{n}$ $^{19}\text{F}(\text{p},\text{pn})^{18}\text{F}$

6   $^{20}_{10}\text{Ne} + ^{2}_{1}\text{H} \rightarrow ^{18}_{9}\text{F} + ^{4}_{2}\text{He}$ $^{20}\text{Ne}(\text{d},\alpha)^{18}\text{F}$

That's a lot of options; but wait, there's more! Figure 7.3 illustrates possible "low-energy" nuclear reactions leading to a particular product. Low-energy simply means these reactions are accessible with fairly conventional accelerators, not huge, high-power atom smashers. Figure 7.3 is laid out just like a chart of the nuclides with numbers of neutrons increasing from left to right and atomic number increasing from bottom to top. The square in the center represents the desired product nuclide, and the reaction types needed to get from a particular starting nuclide to the product are given in the surrounding squares. For example, we know that an n-γ reaction produces a nuclide that is one square to the right on the chart. Sure enough, "n,γ" is listed as a possible reaction type one square to the left of the desired product. A deuterium-proton (d,p) reaction accomplishes the same result as an n-γ reaction and therefore appears in the same square. Adding deuterium (^{2}H) increases both N and Z by one; emitting a proton lowers Z by one—the net result is an increase of N by one.

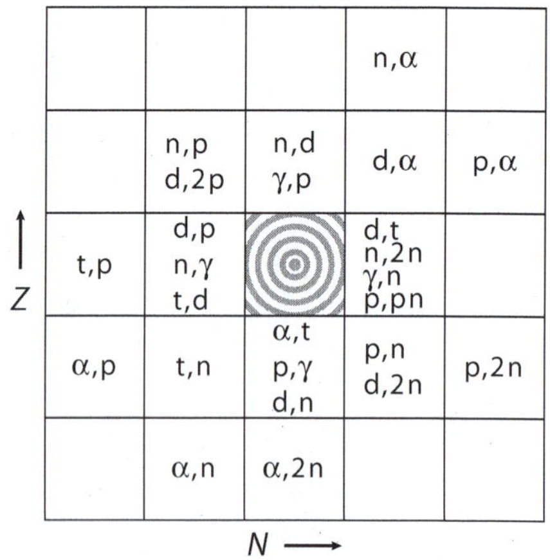

		n,α	
	n,p d,2p	n,d γ,p	d,α	p,α
t,p	d,p n,γ t,d	⦿	d,t n,2n γ,n p,pn	
α,p	t,n	α,t p,γ d,n	p,n d,2n	p,2n
	α,n	α,2n		

FIGURE 7.3   Nuclear reactions leading to a particular product (center square).

Our list of possible reactions leading to ^{18}F suddenly looks a little short. That's because the list above is limited to a single reaction starting from a naturally occurring nuclide. This is our first criterion in considering potential nuclear reactions—it's usually easier to start with a nuclide that is naturally occurring. Let's see if we can narrow the list a bit more. Reactions that emit more than one particle (such as p,pn) generally require more energy than those that only emit one particle. We can cross off Reaction 5.

We also want a high yield of product. Just like chemical reactions, sometimes nuclear reactions work well and make lots of product and other times they don't. We'll discuss this more later. Finally, we'd also like to keep the energy requirements within reason. Of the remaining reactions, numbers 4 and 6 have the highest yields and lowest energy requirements. Deciding between these two may depend on what kind of gun is available—in other words, can you easily accelerate protons (Reaction 4) or deuterium (Reaction 4). It may also depend on how easy it is to separate the product. The advantage of Reaction 6 is that the ^{18}O can be incorporated into water molecules, then the ^{18}F will be produced as the fluoride anion, which can easily be separated from the ^{18}O-water using an ion-exchange column. Reaction 4 is the winner, and is commonly used to produce ^{18}F for PET imaging. Let's check out the energetics of this reaction.

$$
\begin{array}{lll}
\text{Mass of reactants} & {}^{18}\text{O} & 17.999160 \text{ u} \\
 & {}^{1}\text{H} & \underline{+\,1.007825 \text{ u}} \\
 & & 19.006985 \text{ u} \\
 \\
\text{Mass of products} & {}^{18}\text{F} & 18.000938 \text{ u} \\
 & {}^{1}\text{n} & \underline{+\,1.008665 \text{ u}} \\
 & & 19.009603 \text{ u}
\end{array}
$$

$$\text{Reactants}-\text{products} = 19.006985 - 19.009603 = -0.002618 \text{ u}$$

Energy required:
$$Q = \frac{931.5 \text{ MeV}}{\text{u}} \times \left(-0.002618 \text{ u}\right) = -2.44 \text{ MeV}$$

Threshold energy:
$$E_{tr} = 2.44 \times \left(\frac{18+1}{18}\right) = 2.57 \text{ MeV}$$

Effective Coulomb barrier:
$$E_{ecb} \approx 1.11 \times \left(\frac{18+1}{18}\right) \times \left(\frac{8 \times 1}{\sqrt[3]{18} + \sqrt[3]{1}}\right) = 2.59 \text{ MeV}$$

For this reaction the threshold energy and the effective Coulomb barrier are about the same, and they are both low. The effective Coulomb barrier is slightly higher, so the minimum energy required for this reaction to proceed in good yield is 2.59 MeV.

7.2   CROSS SECTION

You've probably noticed that we've always qualified the minimum energy requirement with the phrase "in good yield." This suggests that the reaction can proceed at lower energies, but in poor yield. This is true. Nuclear reactions sometimes do proceed at lower energies. This is called **tunneling**. It is somewhat analogous to a car driving through a hill rather than over it. Driving through will take a lot less energy than driving over. Sadly, tunneling isn't common enough to run nuclear reactions in good yield on the cheap, so it'll still be a good idea to give those projectiles enough energy to ensure success.

Like chemical reactions, you don't always get what you want with nuclear reactions. Various products can be obtained with the same target/projectile combo. For example, hitting an ^{63}Cu target with a proton projectile can result in all of the products shown in figure 7.4. The product ratios will depend on the energy of the projectiles and the probabilities (**cross sections**) for the reactions. With higher-energy projectiles, a greater product distribution is generally observed.

Every possible nuclear reaction has a cross section, which is symbolized by the Greek letter sigma (σ). When the projectile is a neutron, a subscript following σ indicates what is emitted along with the formation of product. For instance, σ_γ is the cross section for an n, γ reaction. While it is best to think of the cross section as the probability of success, it is more strictly defined as the *apparent* cross sectional area of the target nucleus as seen by the projectile. It has units of **barns** (b). One barn is equal to 10^{-24} cm^2 (10^{-28} m^2), which is roughly the projected area of the average nucleus. If nuclear size were the only criterion for the chances of a reaction proceeding, then we'd expect cross section to be low for low-Z nuclides and high for high-Z nuclides. While this is sometimes true, there are other criteria.

Nuclear cross sections depend not only on Z, but also N (and therefore A), the density of the target, and the charge, mass (size), and velocity of the projectile. Figure 7.5 shows how dramatically the probability of all neutron-induced reactions (σ_a) varies with the velocity (energy) of the neutron for the element silver. The general trend is a decreasing probability of reaction as the energy of the neutron increases. This can be rationalized in terms of time. As the neutron moves faster, it spends less time near each nucleus it passes by, thereby decreasing the probability of interaction. Neutrons that have been slowed by room temperature water and/or plastic are called **thermal neutrons**, and have an energy of about 0.025 eV.

Cold neutrons have even lower energies as they are passed through materials below room temperature. For example, cold neutrons are generated at Oak Ridge National Laboratory by passing them through supercritical hydrogen at 20 K. Fast neutrons are those with energies significantly higher than room temperature.

The "spikes" observed in figure 7.5 for fast neutrons indicate that, for neutrons with certain energies, the probability of reaction increases dramatically. These spikes

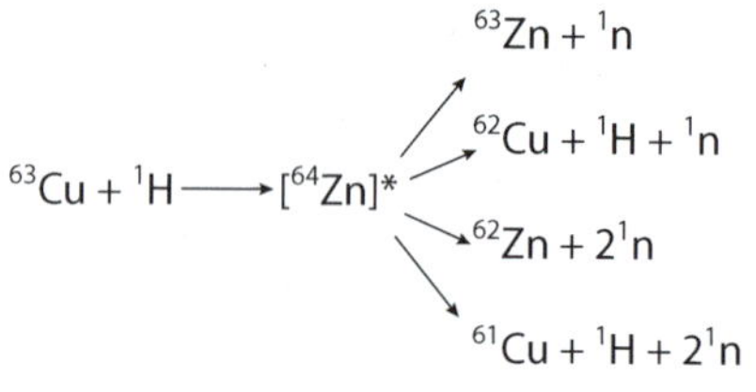

$$^{63}Cu + {}^{1}H \longrightarrow [^{64}Zn]^* \begin{cases} {}^{63}Zn + {}^{1}n \\ {}^{62}Cu + {}^{1}H + {}^{1}n \\ {}^{62}Zn + 2{}^{1}n \\ {}^{61}Cu + {}^{1}H + 2{}^{1}n \end{cases}$$

FIGURE 7.4   Possible products for the reaction of ^{63}Cu with ^{1}H.

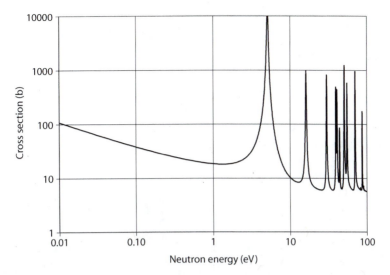

FIGURE 7.5   Probability of a neutron-induced reaction with a silver target as the energy of the neutron increases. Produced using data from the evaluated nuclear data file (ENDF) at http://www.nndc.bnl.gov/.

are due to "**resonance capture**," which is the formation of a compound nucleus with an especially favorable amount of excitation energy. The energy that they occur at is called the **resonance energy**. They usually occur for 1–100 eV neutrons.

What do the spikes mean? If a particular nuclear reaction is desired, then there may be a specific projectile energy that will greatly increase the probability of reaction. Therefore, it may be desirable to run that reaction with projectiles at the resonance energy. The (enhanced) probability that a reaction will occur at a particular resonance energy is called the **resonance integral**.

The bottom line is that variations in nuclear cross sections are not as simple as comparing sizes of target nuclei. They sometimes vary wildly even for closely related target nuclides. For example, the thermal neutron cross section for the n,γ reaction for ^{1}H is 0.332 b, but is 0.52 mb (*milli*barns!) for ^{2}H.

Not every interaction between projectile and target results in a reaction. Sometimes the projectile is simply scattered by the target. The cross section for all possible interactions can be represented as σ_{tot}.

The nature of the product makes a difference because it could be stable or radioactive. If it's hot, it'll be decaying as it is produced, making it a little more challenging to calculate yield.

7.3   YIELD

So, how can we tell how much product is made? Reaction yield depends on how much target there is, the projectile flux, the reaction cross section, time, and the nature of the product. **Flux** (Φ, the Greek letter phi) is a measure of how many projectiles pass through a certain area per unit time. Our focus is on neutron flux, so it'll have units of neutrons per square centimeter per second.

We'll look at three examples, each with a different product scenario: (1) production of a stable product; (2) production of a radioactive product; and (3) production of a radioactive product with a long half-life. In every scenario we'll assume: (1) the flux does not change with time; (2) the number of target atoms (A) is so large, relative to the number of product atoms (B), it can be considered constant throughout the time of irradiation (t); and (3) the target is thin enough that the flux entering the target is not significantly decreased as it goes through. These assumptions may seem a little perilous, but they are necessary to make the math a bit more tractable, and generally speaking, they are true.

When the product is stable, the amount of product B formed can be calculated using the following formula:

$$N_B = N_A \Phi \sigma t \tag{7.5}$$

where N_B is the number of atoms of the product nuclide formed after an irradiation time t with units of seconds. N_A is the number of target atoms A (the target doesn't need to be pure nuclide), Φ is the flux of projectile x in particles per square centimeter per second, and σ is the reaction cross section in square centimeters.

Example: A sample containing 1.00 g of ^{16}O is placed in a nuclear reactor for 10.0 h. The reactor has a neutron flux of 6.00×10^{10} n/s · cm². Assuming that $^{16}O(n, \gamma)^{17}O$ is the only reaction that takes place, how many ^{17}O atoms are formed?

The cross section (0.19 mb) is obtained from the chart of the nuclides and converted to square centimeters. Time also needs to be converted to seconds. Finally, the number of target atoms needs to be calculated from their mass.

$$0.19 \text{ mb} = 0.19 \times 10^{-27} \text{ cm}^2$$

$$10.0 \text{ h} \times \frac{60 \text{ min}}{h} \times \frac{60 \text{ s}}{\min} = 36,000 \text{ s}$$

$$N_A = \left(1.00 \text{ g} \times \frac{\text{mol}}{15.995 \text{ g}} \times \frac{6.022 \times 10^{23} \text{ atom}}{\text{mol}}\right) = 3.76 \times 10^{22} \text{ atom}$$

$$N_B = N_A \Phi \sigma t$$

$$= (3.76 \times 10^{22} \text{ atom}) \times \left(6.00 \times 10^{10} \text{ n/cm}^2 \cdot \text{s}\right)\left(0.19 \times 10^{-27} \text{ cm}^2\right)\left(36,000 \text{ s}\right)$$

$$= 1.55 \times 10^{10} \text{ atoms of } ^{17}O \text{ formed}$$

If the rate of formation of product is desired, the equation above can be modified by dividing both sides by time:

$$\text{rate of formation of B} = N_B/t = N_A \Phi \sigma \tag{7.6}$$

If the product is radioactive, we'll have to worry about the stuff decaying as we make it. The amount of hot B formed after irradiating for time t is given by the following formula:

$$N_B = \frac{\sigma \Phi N_A}{k}\left(1 - e^{-kt}\right) \qquad (7.7)$$

where k is the decay constant, and should have units of per second. The other variables are the same as we've seen before. Note that N_B can be calculated only for the moment the target is removed from the projectile beam. At any time after that, N_B will be smaller due to its radioactive decay.

Since $A = kN$, the equation above can also be written:

$$A_0 = \sigma \Phi N_A\left(1 - e^{-kt}\right) \qquad (7.8)$$

where A_0 is the activity of the product at the moment the target is removed from the projectile beam, and has the units of decays per second (dps). Activity is therefore a function of irradiation time (t) and will eventually **saturate**. Saturation occurs when the rate of formation of product is equal to the rate of decay of product. Sound familiar? It's a bit like nuclear equilibria (Chapter 2). For production of radionuclides, irradiation for seven product half-lives gives 99% of saturation, and is a reasonable time to harvest the product nuclide, if maximum yield and minimum time are desired. The path to saturation is illustrated in figure 7.6. Activity gradually increases as more and more

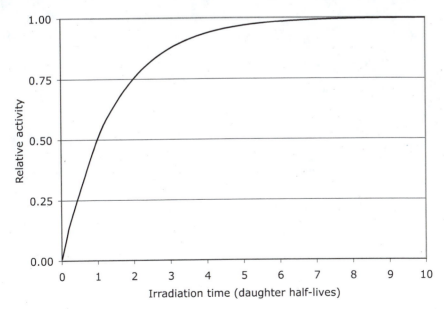

FIGURE 7.6 Activity as a function of irradiation time.

product is made, until it reaches saturation. At saturation, e^{-kt} becomes relatively small and $1 - e^{-kt}$ (sometimes called the saturation term) approaches one. The equation above then simplifies to:

$$A_{sat} = \sigma \Phi N_A \qquad (7.9)$$

at saturation. A_{sat} is the activity at saturation.

The equation for A_0 in eq. 7.8 finally gives us something tangible—activity. Counting atoms (N) is possible, but activity is easy to measure. This equation allows easy experimental determinations of flux, cross section, or the amount of a particular nuclide in a sample. The last is known as **Neutron Activation Analysis**—a powerful analytical tool, especially if the target nuclide has a large neutron cross section. It is capable of detecting as little as 10^{-14} g of target/g of sample.

Example: A 2.50 g sample of ^{18}O water was placed in a cyclotron beam of protons (flux $= 3.01 \times 10^3$ p/s·cm²) for 5.00 hs, resulting in an activity of 153 dps. Assuming that ^{18}O(p, n)^{18}F is the only reaction that takes place, calculate the cross section for this reaction.

First, calculate the number of ^{18}O atoms in the target, remembering it is $H_2{}^{18}O$.

$$2.50 \text{ g } H_2{}^{18}O \times \frac{\text{mol } H_2{}^{18}O}{20.0 \text{ g } H_2{}^{18}O} \times \frac{1 \text{ mol } {}^{18}O}{1 \text{ mol } H_2{}^{18}O} \times \frac{6.022 \times 10^{23} \text{ atom}}{\text{mol}} = 7.52 \times 10^{22} \text{ atom}$$

The half-life of ^{18}F is 1.8295 h, therefore:

$$A_0 = \sigma \Phi N_A \left(1 - e^{-kt}\right)$$

$$153 \text{ dps} = \sigma \times \left(3.01 \times 10^3 \text{ p/s} \cdot \text{cm}^2\right) \times (7.53 \times 10^{22} \text{ atom})\left(1 - e^{-(\ln 2/1.8295 \text{ h}) \times 5.00 \text{ h}}\right)$$

$$\sigma = 7.95 \times 10^{-25} \text{ cm}^2 = 0.795 \text{ b}$$

Note that this reaction had not yet reached saturation. We can see this by dividing the half-life into the irradiation time:

$$5.00 \text{ h}/1.8295 \text{ h} = 2.73 \text{ daughter half-lives}$$

At least seven daughter half-lives need to pass before saturation is reached. The other way to see that saturation has not been reached is to calculate the saturation term separately. If it is very close to one, then saturation has been reached.

$$\left(1 - e^{-kt}\right) = \left(1 - e^{-(\ln 2/1.8295 \text{ h}) \times 5.00 \text{ h}}\right) = 0.850$$

Not close enough.

When the radioactive product has a really long half-life, the math also simplifies. When $t_{1/2} \gg t$, then $(1 - e^{-kt}) \approx kt$. Therefore, our equation for the activity at the time the sample is removed from the beam simplifies to:

$$A_0 = \sigma \, \Phi \, N_A kt \qquad (7.10)$$

Nuclide production is linear with time! Because of the long half-life, more product is made as more time passes. As a result, the issue of saturation is a bit moot if the half-life of the product is too long. Irradiating for seven (or more) daughter half-lives may not be practical.

If a mixture is being irradiated and there are short-lived *and* long-lived nuclides being formed, we can take advantage of their difference in time to saturation. If we want the short-lived nuclide(s), we irradiate for a short time—because very little of the long-lived product will be formed. If we want the long-lived nuclide(s), we irradiate for a long time, take the sample out, and wait for the short-lived stuff to decay.

7.4   ACCELERATORS

How do you get energetic projectiles? Probably not at your local big box discount store. Aside from radioactive decay, bombarding a gas with energetic electrons can produce charged particles. The positive gas ions produced are separated by attraction to an electrode with negative voltage. They are then accelerated through electrical potentials. Passing a particle through a 1000 V potential would give it another kiloelectronvolt per unit charge. There are a couple of ways to get this done.

A schematic for a traditional linear accelerator is shown in figure 7.7a. Linear accelerators are often called linacs, and accelerate charged particles in a straight line through single or multiple stages. Each stage is a cylindrical electrode (aka drift tube) that has the opposite charge of the particle as the particle moves toward it, then the same charge of the particle as it is moving away. Since opposite charges attract and the same charge repels, the particle accelerates as it approaches *and* as it moves away from the drift tube. The tube has no charge while the particle is inside the tube, thus it is coasting (or drifting!) as it passes through the tube. The final energy of the particle beam depends on the voltage applied at each stage, and the total number of stages.

While they can accelerate a variety of charged particles, linacs are most commonly used to accelerate electrons. As the electrons are accelerated, the stages get longer; this is done to ensure the electron spends the same amount of time traveling through each tube. This simplifies the electronic switching between positive and negative charges on the tubes. In order to accelerate particles to very high velocities (energies), this type of linac needs to be very large—some are even a couple miles in length. While this is fine for large research centers, more compact designs are necessary in more common settings such as a hospital.

Most **modern-day linear accelerators** use microwaves to accelerate electrons. These microwaves are often produced in much the same way they are for an ordinary microwave oven. Microwaves, like all forms of electromagnetic radiation, can only travel at the speed of light, so the radiation itself cannot be used—the photons are

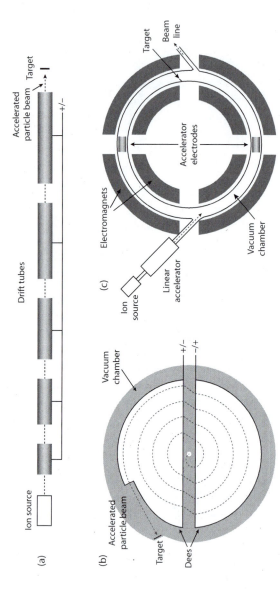

FIGURE 7.7   Schematic representations of (a) linear accelerator, (b) cyclotron, and (c) synchrotron.

not going to slow down and push the particles to higher and higher speeds. Instead it is the phase of these waves can be manipulated to accelerate electrons.

Remember that electromagnetic radiation behaves like waves in many respects. We can think of phase in relation to the peaks and troughs of a wave. A peak represents one phase, and the trough represents the opposite phase. The electron is pushed and pulled by these phases, so if we arrange them just right, they will accelerate the electron. It's a bit like a surfer riding a wave as it approaches the beach. The surfer is constantly moving downhill, and is being pushed forward by the wave. In our case the wave is the phase and it's accelerating.

There's another way to look at the acceleration of charged particles in this kind of linear accelerator. As the microwave moves through a metal tube, it will induce temporary positive and negative electrical charges in the walls of the tube. If these temporary charges move just right, they can be used to accelerate charged particles.

For clinical applications, such as radiation therapy, these accelerators only need to be a couple of feet long, and can accelerate electrons up to 20 MeV. The Stanford linear accelerator uses microwaves to accelerate electrons up to 50 GeV and is nearly two miles long!

Cyclotrons are the most commonly used accelerators for radionuclide production in hospitals and other, more ordinary places. They take advantage of the fact that charged particles will float in a circle when placed in a constant magnetic field, and that the radius of the circle is directly related to the velocity (energy) of the particle. In other words, the faster the particle is moving, the bigger the circle it'll make. The size of the circle also depends on the strength of the magnetic field—the stronger the field, the smaller the circle.

A schematic for a simple cyclotron is illustrated in figure 7.7b, although the magnets have been removed for clarity. Only the accelerator portion of the cyclotron is shown. Imagine two round magnets sitting above and below the page in figure 7.7b—just like an Oreo™[1] cookie, where the magnets are the crunchy chocolate cookie, and the particles are accelerated in the delicious creamy filling.

Charged particles are injected into the center of the cyclotron (the small white circle), and immediately begin to move in a circle parallel to the two magnets. As it circles around, it moves in and out of two hollow, semicircular electrodes. These electrodes are called **dees** because they look like the letter *D*. Imagine making them by cutting a tuna can in half along the can's diameter and taking all the tuna out (somewhere, a cat is purring...). As is true for all particle accelerators, the cyclotron is kept under a vacuum so that the accelerating particles don't run into any matter along the way.

Just like the traditional linear accelerator described above, acceleration of the charged particle takes place as it travels between the dees. As it leaves one dee, that dee is given the same charge as the particle, pushing it away. At the same time, the dee the particle is approaching is given the opposite charge, attracting the particle to it. As it's accelerated, the particle moves in a larger circle, giving it a kind of spiral path as it is accelerated in the cyclotron. It is not a smooth spiral, as the particle is not accelerated as it travels through the dees and therefore moves along a semicircular

---

[1] Oreo is a registered trademark of Kraft Foods Inc., Northfield, Illinois.

path with a fixed radius. Eventually, the circular path of the particle is large enough (energy is high enough) to allow it to exit the cyclotron.

Cyclotrons are not very good at accelerating electrons. Because of their low mass, it doesn't take a whole lot of energy ($\sim$100 keV) to get them traveling at velocities approaching the speed of light. When particles approach the speed of light, increasing amounts of energy are required to accelerate the particle by increasingly small amounts (Chapter 5). In a cyclotron, this means that once the particle gets to a certain velocity, it'll be difficult to continue to accelerate it, and it will pretty much stay in a circle (not continuing to spiral). Additionally, there are practical limits to the strengths of the electromagnetic fields within the cyclotron. As a result, cyclotrons tend to have energy limits for various particles. For most particles accelerated in cyclotrons (^{1}H, ^{2}H, ^{4}He, etc.) this limit is 25–50 MeV.

A common nuclide produced by cyclotrons for clinical use is ^{18}F. As we've already seen, the reaction below requires a minimum energy of $\sim$2.6 MeV. Protons can easily be accelerated in cyclotrons to this energy.

$$^{18}_{8}O + {}^{1}_{1}p \rightarrow {}^{18}_{9}F + {}^{1}_{0}n$$

Products of cyclotron reactions tend to be proton-rich. Therefore, they tend to decay via β^+ emission or electron capture. Atomic number (Z) almost always changes in these reactions, making it easier to isolate isotope-free product because the target and the product will have differing chemical properties. Cyclotron reactions are even less efficient than neutron activation. This makes sense because we are now trying to get two positively charged particles to hit each other. Coulomb repulsion forces must be overcome for this to happen. As a result, cyclotrons produce nuclides in pretty low yields, and they tend to be a little pricey.

Table 7.1 shows some of the more common nuclides produced by cyclotrons. Note that ^{11}C, ^{13}N, and ^{15}O can all be produced by cyclotron reactions. These nuclides all have short half-lives (≤ 20 min), so they must be produced on-site and quickly incorporated into a drug, then delivered to the patient. These elements are all important in biological molecules—if we could place one of these nuclides in a biomolecule, then it would have all the same chemical and biological properties, but the radioactive label would allow us to see where that molecule goes in the body. This is an exciting area of current research in nuclear medicine, with the potential to develop a number of highly effective radiopharmaceuticals.

Another type of circular accelerator that can accelerate particles up to incredibly high energies (TeV $= 10^{12}$ eV) are **synchrotrons**. Synchrotrons use oscillating magnetic and electrical fields to accelerate particles on a fixed radius. As the particle energy increases, so does the strength of the magnetic field, keeping the particles moving through the same circle. Synchrotrons are enormous. The Large Hadron Collider near Geneva, Switzerland, is over 27 km in circumference, and accelerates protons to 7 TeV.

7.5 COSMOGENIC NUCLIDES

Just about all of the background radiation detected on the surface of the planet originates from cosmic radiation. As depicted in figure 7.8, cosmic radiation originates

TABLE 7.1
Some Cyclotron-Produced Radionuclides Used in Nuclear Medicine

Product Nuclide	Decay Mode	Common Production Reaction(s)	Natural Abundance of Target Nuclide (%)
		$^{14}N(p,\alpha)^{11}C$	99.63
^{11}C	β^+, EC	$^{10}B(d,n)^{11}C$	19.9
		$^{11}B(p,n)^{11}C$	80.1
^{13}N	β^+	$^{16}O(p,\alpha)^{13}N$	99.76
		$^{12}C(d,n)^{13}N$	98.93
^{15}O	β^+	$^{14}N(d,n)^{15}O$	99.63
		$^{15}N(p,n)^{15}O$	0.368
^{18}F	β^+, EC	$^{18}O(p,n)^{18}F$	0.205
		$^{20}Ne(d,\alpha)^{18}F$	90.48
^{22}Na	β^+, EC	$^{23}Na(p,2n)^{22}Na$	100
^{43}K	β^-	$^{40}Ar(\alpha,p)^{43}K$	99.60
^{67}Ga	(EC,γ)	$^{68}Zn(p,2n)^{67}Ga$	18.75
^{111}In	(EC,γ)	$^{109}Ag(\alpha,2n)^{111}In$	48.16
		$^{111}Cd(p,n)^{111}In,$	12.49
^{123}I	(EC,γ)	$^{122}Te(d,n)^{123}I$	2.55
		$^{124}Te(p,2n)^{123}I$	4.74
^{201}Tl	(EC,γ)	$^{201}Hg(d,2n)^{201}Tl$	13.18

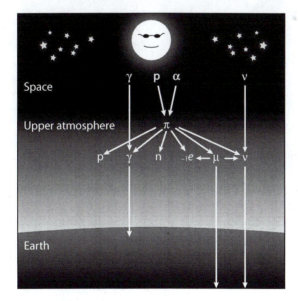

FIGURE 7.8 Cosmic radiation.

from our sun, other stars in our galaxy, and other galaxies. These stars emit mostly light nuclei (solar wind) and neutrinos. About 90% of the particles bombarding our planet are *very*-high-energy (up to 10^{10} GeV) protons and alpha particles. When these screaming nuclear projectiles slam into the atmosphere, they annihilate themselves, each one producing dozens of pions (π). Pions are subatomic particles that are commonly produced in high-energy (>400 MeV) nuclear reactions. In the average nucleus, pions are exchanged between nucleons and thereby hold the nucleus together—somewhat like the way electrons hold atoms together in molecules.

Pions don't last long on their own ($t_{1/2} \approx 10^{-6}$ s). They decay to neutrinos and muons (μ). Muons are yet another subatomic particle and are similar to electrons. Fifty to eighty percent of the muons formed actually make it to the earth's surface. Muons don't last long either ($t_{1/2} = 2 \times 10^{-6}$ s), decaying to electrons and neutrinos. Even so, much of the cosmic radiation that reaches the surface of our planet is in the form of muons. High-energy photons (γ) and neutrinos (μ) make up most of the rest of the cosmic radiation flux at the planet's surface.

Thanks mostly to the Earth's magnetic field, and to all the matter in our atmosphere, we are shielded from all but about 5% of the cosmic radiation that bombards our planet. The significantly higher radiation flux in space is an important concern for astronauts spending considerable amounts of time there, and is a major hurdle to long space missions such as travel to other planets both within and outside of our solar system.

Pions also collide with matter in the atmosphere producing electrons, neutrons, protons, and photons. Some of these particles can initiate nuclear reactions forming ^{3}H, ^{10}B, and ^{14}C, providing a constant source of these (and other!) radionuclides. As mentioned in Chapter 2, ^{14}C has a half-life of 5715 years, is a soft β^- emitter ($E_{max} = 158$ keV), and is the nuclide of interest when performing carbon dating. ^{14}C is formed by the ^{14}N(n,p)^{14}C reaction when a thermal neutron is involved. If the neutron is more energetic (fast), formation of ^{12}C is preferred via ^{14}N(n,t)^{12}C.

QUESTIONS

1. Complete the following and write balanced equations ($A + x \rightarrow B + y$).

^{18}O(n, β^-)	^{14}N(α, p)	^{141}Pr(γ, 2n)
^{238}U(α, n)	^{59}Fe(α, β^-)	^{10}B(d, n)

2. Write out the neutron capture reaction for ^{191}Ir. How much energy would be produced by this reaction?
3. Write out the α,p reaction for ^{12}C. What is the compound nucleus formed in this reaction? If the alpha particle has an energy of 7.69 MeV (from the decay of ^{214}Po), what is the excitation energy of the compound nucleus?
4. What is the minimum amount of energy required for the reaction in Question 3 to proceed in good yield?
5. Calculate the velocity (m/s) and wavelength (nm) of a thermal neutron.
6. The reaction below occurs in a small cyclotron. Calculate the effective Coulomb barrier for this reaction, the amount of energy produced or required

by this reaction (Q), and the kinetic energy of the tritium ion necessary to make this reaction proceed in good yield.

$$^{3}_{1}H + ^{124}_{50}Sn \rightarrow ^{126}_{51}Sb + ^{1}n$$

7. ^{111}In is a nuclide used in clinical settings for diagnostic scans. It can be made using an α,2n reaction. How much energy is required for this reaction to proceed in good yield? Could this reaction be performed using a cyclotron? Briefly explain.

8. A thin sample of Mn (549 mg) was placed in a neutron howitzer with a flux of 1.50×10^4 n/cm$^2 \cdot$s for 5.00 h. Calculate the activity of the sample when it was removed from the howitzer. If a product is formed, what is it, and how much (mg) is made?

9. A 0.486 g sample of gold was placed in a cyclotron beam of helium ions (flux $= 1.03 \times 10^4$ ions/s $\cdot$ cm^2) for 67 h, resulting in an activity of 97 dps. Assuming that the only particle emitted from the compound nucleus is ^{3}He, calculate the cross section (b) for the reaction and the number and identity of the atoms produced. Is there a simpler way to perform this reaction? Explain.

10. Give at least three transmutation reactions that could produce ^{15}O (used for PET). On the basis of threshold energy and effective Coulomb barrier, which would be the best one to use?

11. A neutron howitzer employs 25 gal. of water for shielding and has a flux of 4.9×10^4 n/cm$^2 \cdot$s at the neutron source. If the source is a cylinder that is 4.0 in. tall and 1.0 in. in diameter, what percentage of the water will be DOH (D$=$deuterium) after one year? Would you expect a significant amount of TOH (T$=$tritium) or D$_2$O to form in this time? Briefly explain.

12. Cold fusion enthusiasts recently reported that the ^{133}Cs(d, γ)^{135}Ba reaction takes place close to room temperature. As the science editor for a prestigious journal, you are asked to review their work. How would you respond?

13. A student places a sample containing an unknown element in a neutron howitzer for 1.5 h. Upon removal, the student observes that a nuclide was formed with a half-life of about 3.8 h, but the count rates were low. How long should the student keep the sample in the howitzer to achieve maximum activity?

14. It has been reported that the reaction of ^{48}Ca with ^{186}W produces ^{40}Mg. If only one other nuclide is formed by this reaction, write out a balanced nuclear reaction. What might be the driving force for the formation of these particular products?

8 Fission and Fusion

Nuclear fission and fusion are special types of nuclear reactions. Because of their prominence in modern science and society, they will both be discussed in some detail in this chapter.

8.1 SPONTANEOUS FISSION

Spontaneous fission is a form of decay where a nuclide splits into two relatively large chunks. It is only observed for the heavier nuclides, and even then is only observed as a primary decay mode for a handful of nuclides. Spontaneous fission can be generically represented by:

$$^A_Z X \rightarrow\ ^{A_1}_{Z_1} L +\ ^{A_2}_{Z_2} M + \nu n$$

where L and M are **fission products** (or fragments), and νn indicates the release of a small whole number (typically 0–4) of neutrons. A wide variety of fission products are typically formed by the decay of a particular parent, but products with magic numbers (like 50 and 82) of neutrons and or protons have increased odds of formation.

The only naturally occurring nuclide that undergoes spontaneous fission is ^{238}U, however its branch ratio for this form of decay is quite low (0.00005%). ^{238}U has 92 protons and 146 neutrons, and is therefore an even–even (*ee*) nuclide. Even–even nuclides are somewhat more likely to undergo fission than nuclides with odd numbers of protons and/or neutrons.

The neutrons released in fission reactions usually escape their immediate surroundings and can interact with material that is a significant distance away. The fission products formed often do not travel far. These products have a significant amount of kinetic energy, which is mostly dissipated as heat. The kinetic energies of the fission products are high because of the strong Coulomb repulsion of the two fragments right after they separate. The total kinetic energy (K_T) in MeV can be estimated using the following formula:

$$K_T = 0.80 \times \frac{Z_1 Z_2}{\sqrt[3]{A_1} + \sqrt[3]{A_2}} \tag{8.1}$$

where Z_1 and Z_2 are the atomic numbers of the two fission products, and A_1 and A_2 are their mass numbers. Notice that this formula is similar to the formula for calculating the Coulomb barrier (Chapter 7). The only difference is the constant value—it was 1.11 for the Coulomb barrier and is 0.80 here. The difference lies in the distance between the centers of the nuclei. When calculating Coulomb barrier, we assume the nuclei are spheres that are just barely touching each other. When determining kinetic energy following fission, it is better to assume some separation between the spheres, as this gives more realistic results (fig. 8.1). The greater the separation, the lower the energy should be.

The kinetic energies of the individual fragments can then be estimated using the same conservation of momentum formulas used to approximate recoil and alpha particle energies in alpha decay (Chapter 4).

$$K_1 = K_T\left(\frac{A_2}{A_1 + A_2}\right) \tag{8.2}$$

$$K_2 = K_T\left(\frac{A_1}{A_1 + A_2}\right) \tag{8.3}$$

Some of the variables have changed, but these are the same formulas as equations 4.1 & 4.2.

Example: Calculate the total energy released and estimate the kinetic energies of the fission products in the spontaneous fission reaction below.

$$^{238}_{92}U \rightarrow {}^{134}_{52}Te + {}^{102}_{40}Zr + 2{}^{1}_{0}n$$

Note that these fission products are just two of a great many possible spontaneous fission products for ^{238}U. Notice also that ^{134}Te has a magic number (82) of neutrons and therefore can be expected to be formed in somewhat higher yield. To find the total energy released by this decay, we need to determine how much mass is lost.

Mass of reactant	^{238}U	238.050783 u
Mass of products	^{134}Te	− 133.911540 u
	^{102}Zr	− 101.922981 u
$2n = 2(1.008665\ u) =$		− 2.017330 u
Mass lost		0.19893 u

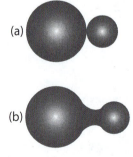

(a)

(b)

FIGURE 8.1   The differences in nuclear separation for (a) collision and (b) fission.

Total energy of decay:

$$E = \frac{931.5 \text{ MeV}}{\text{u}} \times 0.19893 \text{ u} = 185.3 \text{ MeV}$$

Now, let's estimate the kinetic energy of the two fragments.

$$K_\text{T} = 0.80 \times \frac{Z_\text{Te} Z_\text{Zr}}{\sqrt[3]{A_\text{Te}} + \sqrt[3]{A_\text{Zr}}} = 0.80 \times \frac{52 \times 40}{\sqrt[3]{134} + \sqrt[3]{102}} = 170 \text{ MeV}$$

$$K_\text{Te} = K_\text{T} \left(\frac{A_\text{Zr}}{A_\text{Te} + A_\text{Zr}} \right) = 170 \text{ MeV} \times \frac{102}{134 + 102} = 73 \text{ MeV}$$

$$K_\text{Zr} = K_\text{T} \left(\frac{A_\text{Te}}{A_\text{Te} + A_\text{Zr}} \right) = 170 \text{ MeV} \times \frac{134}{134 + 102} = 97 \text{ MeV}$$

These formulas provide an estimate only, so it's a good idea to keep the number of significant figures low. The heavier fragment has less energy than the light one—just like alpha decay, only now the difference is a lot less dramatic. Note that the sum of the two fragment kinetic energies equals the total kinetic energy.

$$73 \text{ MeV} + 97 \text{ MeV} = 170 \text{ MeV}$$

Notice also that the total energy of decay is significantly larger than the kinetic energy of the fragments. The remaining 15 MeV (185 MeV–170 MeV) is divided among the neutrons and γ-rays emitted as part of the fission reaction, and the excitation energy of the products. The γ-rays emitted as part of the fission reaction are referred to as **prompt gamma photons**, and are not typically shown when writing out the decay equation.

Because of the low branch ratio, relatively few ^{238}U atoms undergo spontaneous fission. Generally speaking, the odds of spontaneous fission increase with mass number (A). For example, ^{256}Fm decays via spontaneous fission in 92% of all decays. This makes intuitive sense; the bigger the nucleus, the more likely it is to split into two fragments.

8.2 NEUTRON-INDUCED FISSION

If heavy *ee* nuclides are more likely to undergo spontaneous fission, can we get a heavy *eo* nuclide to split if we fire neutrons at it? You bet! In fact, this is a pretty handy way to get fission to occur, rather than sitting around waiting for a nuclide to undergo spontaneous fission. The advantage of adding a neutron to an *eo* nuclide is that we obtain an *ee* nuclide with some excitation energy. Neutron-induced fission reactions (typically referred to simply as "fission reactions") can be generically represented as:

$$A + n \rightarrow [C]^* \rightarrow B + D + \nu n$$

where A is the heavy nuclide, aka the **fissile material**. The compound nucleus is [C]*, and B and D are the fission products. Neutrons are typically released by fission reactions, which is pretty handy if you've got a lot of A and want them to undergo fission. Such a process is known as a **fission chain reaction** and will be discussed in more detail in Chapter 9. Just as in spontaneous fission, the symbol v represents a small whole number, typically 0–4.

Fission reactions tend to be pretty messy—hundreds of different nuclides have been identified as fission products for ^{235}U. These products tend to be rather neutron rich since they originate from a nuclide (^{235}U) with an N/Z ratio that is relatively high when compared to the stable isobars of the products. One example of a fission reaction for ^{235}U is:

$$^{235}_{92}U_{143} + {}^1_0 n \rightarrow {}^{94}_{37}Rb_{57} + {}^{140}_{55}Cs_{85} + 2\,{}^1_0 n$$

$$N/Z \rightarrow 1.55 \qquad\qquad 1.54 \qquad 1.54$$

The N/Z ratio for each nuclide is given directly below it. Despite the emission of two neutrons in this particular fission reaction, the N/Z ratios of the products are pretty much the same as the target nuclide. However, we know (Chapter 1) that nuclides with lower mass numbers will only be stable with lower N/Z ratios. For example, the only stable nuclides for $A = 94$ are $^{94}_{42}Mo_{52}$ ($N/Z = 1.24$) and $^{94}_{40}Zr_{54}$ ($N/Z = 1.35$). Since fission products tend to be neutron rich, they will generally undergo β^- decay. In fact, each fission product can form a short (isobaric) decay series on the way to stability. The series formed by ^{94}Rb consists of ^{94}Sr and ^{94}Y before forming stable ^{94}Zr.

$$^{94}_{37}Rb \xrightarrow{\beta^-} {}^{94}_{38}Sr \xrightarrow{\beta^-} {}^{94}_{39}Y \xrightarrow{\beta^-} {}^{94}_{40}Zr$$

The percentage each nuclide is formed is known as the **fission yield**. It is sometimes difficult to determine the yield of a particular fission product (**independent yield**), because it may have been formed through beta decay of an isobar. Some of these products formed through beta decay series have very short half-lives, making it difficult to run a fission reaction for an amount of time, then analyze the products and get reliable independent yields. Since the products generally retain the same mass number, yields can be determined for all isobars (**cumulative yield**).

The cumulative yields for thermal neutron-induced fission of ^{235}U are collectively graphed in figure 8.2. Fission products for this reaction distribute themselves into two large peaks. Products with mass numbers of 90–100 u and 130–145 u are formed in the highest yields, and are therefore preferred. Generally speaking, the mass numbers from these two peaks add up to the mass number of ^{235}U, meaning that nuclides with differing mass numbers are usually formed. Thermal neutron fission of ^{235}U is therefore **asymmetric**.

Symmetric fission would be the division of the ^{236}U compound nucleus (formed by the interaction of a neutron with ^{235}U) into identical fragments. This clearly does take place, but at a much lower yield than asymmetric fission. This is indicated in

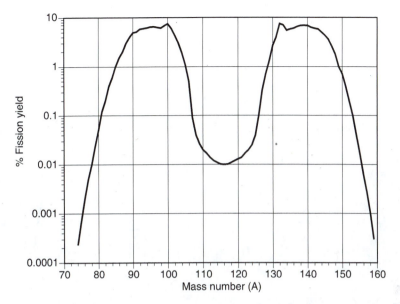

FIGURE 8.2 Fission product yields for thermal neutron fission of ^{235}U.

the middle of figure 8.2 by the formation of nuclides with mass numbers of approximately 118.

$$^{235}_{92}U + ^{1}_{0}n \rightarrow 2\,^{118}_{46}Pd$$

Why is asymmetric fission so strongly preferred? Look closely at the small spike in fission yield at $A = 132$. Something is definitely special about formation of nuclides with this mass number. In particular, $^{132}_{50}Sn_{82}$ is formed. ^{132}Sn has magic numbers of both neutrons and protons, and therefore has greater stability. Fission reactions prefer to form at least one product with a magic number of protons and/or neutrons because of the stability of those nuclides. Many of the other nuclides formed under the peak on the right of figure 8.2 also have either 50 protons (tin) or 82 neutrons. Some of the nuclides under the left peak are preferentially formed because they have 50 neutrons. The rest of the graph is a mirror to these preferences. For example, the small spike at $A = 100$ reflects the spike at $A = 132$, and corresponds to a reaction like the one below:

$$^{235}_{92}U + ^{1}_{0}n \rightarrow ^{132}_{50}Sn + ^{100}_{42}Mo + 4\,^{1}_{0}n$$

^{100}Mo is simply what's left over after the favorable ^{132}Sn is formed and four neutrons are emitted.

The energy of the incident neutron will have an effect on the distribution of the fission products. As shown in figure 8.3, higher-energy (more than thermal) neutrons will increase the probability of symmetric fission relative to the probability of asymmetric fission for ^{235}U. The valley between the two peaks fills in as neutron energy increases. This can be understood in terms of the excitation energy of the compound

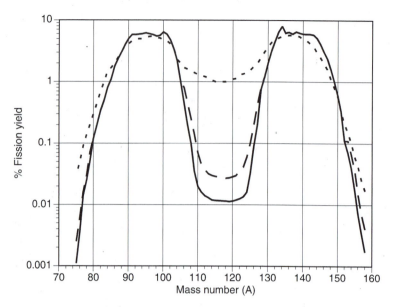

FIGURE 8.3 Fission product distributions for ^{235}U with thermal neutrons (solid line), ~2 MeV neutrons (larger dashed line), and 14 MeV neutrons (smaller dashes). (Data from Flynn, K.F. and Glendenium, L E., Rep. ANL-7749, Argonne National Laboratory, Argonne, IL, 1970.)

nucleus ($[^{236}$U]*). The more energy the neutron has, the greater the excitation energy of the compound nucleus. The greater the excitation energy, the less fussy the compound nucleus will be about forming products with magic numbers of nucleons. Formation of nuclides with magic numbers generates more energy for the reaction; if the reaction already has some extra energy, it isn't as important to generate more.

The fission product distribution also depends on the nuclide undergoing fission. Figure 8.4 shows how the distribution varies between ^{233}U and ^{239}Pu. Notice that the distribution for the lighter of these two nuclides (^{233}U—solid line) is shifted to lower masses, while it is shifted to higher masses for ^{239}Pu (dashed line), when compared to figure 8.2.

Not all heavy *eo* nuclides undergo thermal neutron fission, and not all heavy *ee* nuclides decay via spontaneous fission. Why not? There is an energy barrier to fission. We can think of it as the same as the Coulomb barrier discussed in Chapter 7. Just as there is an energetic barrier to bring two positively charged nuclei together, there is also a barrier to splitting them up. It is called the **fission barrier**, and can be thought of as disruption of the strong force, which holds all the nucleons together in the nucleus. In light of this barrier, it is surprising that spontaneous fission takes place at all. The only way it can take place is by avoiding the barrier. It is generally believed that spontaneous fission takes place via tunneling—the nuclide finds a way through the barrier rather than over it. As you will recall, tunneling is a path less taken, therefore spontaneous fission is a relatively rare form of decay, but can sometimes occur for some of the heaviest known nuclides.

The binding of an additional neutron to a target nucleus can provide the energy to overcome the fission barrier. Let's take a look at some relevant information for

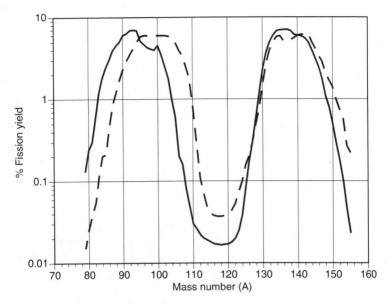

FIGURE 8.4 Thermal neutron fission product distributions for ^{233}U (solid) and ^{239}Pu (dashed).

TABLE 8.1

Relative Tendencies to Undergo Neutron-Induced Fission for Selected Heavy Nuclides

Nuclide	σ_f (b)	σ_γ (b)	Binding E of a Thermal ^{1}n (MeV)	Fission Barrier (MeV)
^{232}Th	0.000003	737	4.8	7.5
^{233}U	530	48	6.8	6.0
^{235}U	586	99	6.5	5.7
^{238}U	0.000003	2.7	4.8	5.8
^{239}Pu	752	269	6.5	5.0

some common nuclides of the actinide series (table 8.1). As discussed in Chapter 7, σ_γ is the cross section (probability) for an n,γ reaction; σ_f is the cross section for a neutron-induced fission reaction. Both cross-sectional values are given in barns to allow easy comparison. For example, ^{232}Th is *much* more likely to form ^{233}Th when hit by a neutron, while ^{233}U is more likely to undergo fission.

For each nuclide in table 8.1, the energy produced by binding a thermal neutron is calculated. This is the same as the excitation energy of the compound nucleus formed by neutron capture. Let's use ^{235}U as an example:

$$^{235}U + {}^1n \rightarrow [{}^{236}U]^*$$

$$Q = \frac{931.5 \text{ MeV}}{\text{u}} \times (235.043923 \text{ u} + 1.008665 \text{ u} - 236.045562 \text{ u})$$

$$= 6.54 \text{ MeV}$$

You may recall that calculation of the excitation energy of a compound nucleus also needs to take into account the kinetic energy of the compound nucleus (E_{KC}—equation (7.4)).

$$E_{KC} \approx E_{KX} \times \left(\frac{A_X}{A_C} \right)$$

The kinetic energy of a thermal neutron is rather low—only 0.0253 eV ($= E_{KX}$)—therefore, the kinetic energy of the compound nucleus formed after thermal neutron capture will be very small relative to the excitation energy. The bottom line is that excitation energy is essentially equal to Q for thermal neutron capture.

Notice that the excitation energy is lower (4.8 MeV) for $^{232}_{90}\text{Th}_{142}$ and $^{238}_{92}\text{U}_{146}$ while it is higher (6.5–6.8 MeV) for $^{233}_{92}\text{U}_{141}$, $^{235}_{92}\text{U}_{143}$, and $^{239}_{94}\text{Pu}_{145}$. This difference arises from the number of neutrons in each nuclide. ^{232}Th and ^{238}U have an even number of neutrons, so addition of another neutron would result in the formation of a nuclide with an odd number of neutrons. ^{233}U, ^{235}U, and ^{239}Pu all start with an odd number of neutrons, and adding another neutron forms a nuclide with an even N. Remember that even numbers of nuclides mean greater stability, and odd numbers mean less stability (Chapter 1). An appropriate analogy is a rock rolling down a hill into a ravine. Generally speaking, the taller the hill and the deeper the ravine, the more energy the rock will have when it hits bottom. Going from even N to odd N (^{232}Th and ^{238}U) means the hill is lower, and the ravine shallower, therefore less excitation energy is produced.

The fission barrier for each nuclide is given in the final column of table 8.1. Those nuclides that generate enough excitation energy from thermal neutron capture to overcome the fission barrier are those that tend to undergo fission rather than an n, γ reaction. Notice that those nuclides with enough energy to overcome the fission barrier also have relatively high σ_f values. Our example nuclide, ^{235}U, looks pretty good for fission. Its excitation energy is 6.5 MeV, but its fission barrier is only 5.7 MeV. As we might then expect, its cross section for a thermal neutron-induced fission reaction is high (586 b!) relative to its cross section toward a n, γ reaction (99 b). For ^{238}U, the opposite is true. Its excitation energy is too low (4.8 MeV) to overcome its fission barrier (5.8 MeV), and it has a very low σ_f (0.000003 b!) relative to its σ_γ (2.7 b). We would therefore expect ^{238}U to predominately form ^{239}U when exposed to thermal neutrons.

The relative cross sections change with the energy of the neutron. Fast neutrons will dramatically increase σ_f while decreasing σ_γ, causing more ^{238}U to fission rather than just form ^{239}U. Again, this can be understood in terms of excitation energy of the compound nucleus. If the neutron has more than about 1 MeV of energy, the excitation energy of the compound nucleus will be greater than the fission barrier, allowing more ^{238}U fission reactions to take place. Therefore, fission is not restricted to excited *ee* nuclides, but to any large nucleus with sufficient excitation energy to overcome its fission barrier.

The energetics of a neutron-induced fission reaction can be handled the same way as for spontaneous fission. Don't forget there's now a neutron on the left side of the arrow. You can save a step in the math by canceling it with one of the product neutrons, as shown in the example below.

Example: Calculate the total energy of the following fission reaction. Estimate the kinetic energy of the fission products.

$$^{235}_{92}U + ^{1}_{0}n \rightarrow ^{141}_{56}Ba + ^{92}_{36}Kr + 3\,^{1}_{0}n$$

Remember, this is just one of hundreds of possible reactions for ^{235}U fission. Following the example for spontaneous fission:

Mass of reactant	^{235}U	235.043923 u
Mass of products	^{141}Ba	− 140.914406 u
	^{92}Kr	− 91.926153 u
net 2n = 2(1.008665 u) =		− 2.017330 u
Mass lost		0.186034 u

Total Energy of decay:

$$E = \frac{931.5\,\text{MeV}}{u} \times 0.186034\,u = 173.3\,\text{MeV}$$

$$K_T = 0.80 \times \frac{Z_{Ba}Z_{Kr}}{\sqrt[3]{A_{Ba}} + \sqrt[3]{A_{Kr}}} = 0.80 \times \frac{56 \times 36}{\sqrt[3]{141} + \sqrt[3]{92}} = 166\,\text{MeV}$$

$$K_{Ba} = K_T\left(\frac{A_{Kr}}{A_{Ba} + A_{Kr}}\right) = 166\,\text{MeV} \times \frac{92}{141 + 92} = 66\,\text{MeV}$$

$$K_{Kr} = K_T\left(\frac{A_{Ba}}{A_{Ba} + A_{Kr}}\right) = 166\,\text{MeV} \times \frac{141}{141 + 92} = 100\,\text{MeV}$$

Just as we saw with spontaneous fission, the total energy of decay is carved up into kinetic and excitation energy of the fragments, neutron energy (velocity), and prompt gamma photons. In this example, only about 7 MeV separate the total energy of decay and K_T. What about the neutrons, prompt γ-rays, and excitation energy? Keep in mind: (1) The total kinetic energy is estimated—it could be a little bit off, and (2) the energy values for the emitted neutrons, prompt γ-rays, and excitation are average values. The math doesn't work perfectly, but it does give us an idea of where all the energy goes.

The fission products in the example above will both undergo decay. The energy released is called **delayed energy**, as opposed to the **prompt energy** associated with the act of splitting the nucleus. The average values for both are given in table 8.2.

TABLE 8.2

Approximate Average Energy Values (MeV) for Thermal Neutron Fission of ^{235}U

Prompt energy	Fission products	167
	Neutrons	5
	Prompt γ-rays	6
Total prompt energy		178
Delayed energy	Beta particles	8
	Neutrino particles	12
	γ-rays	6
Total delayed energy		26
Total fission energy		204

The total kinetic energy for ^{235}U fission products averages about 167 MeV. ^{235}U emits an average of 2.4 neutrons per fission, each carrying about 2 MeV of energy, giving approximately 5 MeV of energy associated with the prompt neutrons. Finally, an average of six γ-rays are emitted per fission of ^{235}U, each packing about 1 MeV of energy.

Since delayed energy is all about the beta decay of the fission products, it is composed of the energy of the beta and neutrino particles and the gamma photons emitted during decay(s) to a stable nuclide. The 6 MeV listed for the delayed γ-rays consist of gamma emissions that are part of the beta decays and are emitted during de-excitation of the fission fragments. On average, the thermal neutron fission products of ^{235}U emit about 26 MeV as delayed energy. This brings the total average energy per fission to about 200 MeV. Let's see how our example compares.

Example: Estimate the delayed energy emitted by the fission reaction given in the previous example.

All we need to do is look at the two nuclides produced and add up the total energy of their isobaric decays to a stable nuclide. Data below are rounded to the nearest tenth million electron volt because of the qualitative nature of this calculation.

$$^{141}_{56}Ba \xrightarrow[3.2\ MeV]{\beta^-} {}^{141}_{57}La \xrightarrow[2.5\ MeV]{\beta^-} {}^{141}_{58}Ce \xrightarrow[0.6\ MeV]{\beta^-} {}^{141}_{59}Pr$$

$$^{92}_{36}Kr \xrightarrow[6.0\ MeV]{\beta^-} {}^{92}_{37}Rb \xrightarrow[8.1\ MeV]{\beta^-} {}^{92}_{38}Sr \xrightarrow[1.9\ MeV]{\beta^-} {}^{92}_{39}Y \xrightarrow[3.6\ MeV]{\beta^-} {}^{92}_{39}Zr$$

^{141}Ba: $3.2\ MeV + 2.5\ MeV + 0.6\ MeV = 6.3\ MeV$

^{92}Kr: $6.0\ MeV + 8.1\ MeV + 1.9\ MeV + 3.6\ MeV = 19.6\ MeV$

Total delayed energy: $6.3\ MeV + 19.6\ MeV = 25.9\ MeV$

Almost spot on!

Some of the fission products are so neutron-rich they sometimes emit delayed neutrons. Don't confuse delayed neutrons with delayed energy. As mentioned at the end of Chapter 4, particles are sometimes emitted from an excited daughter and are called delayed because they are emitted *after* the original decay has taken place. For example, 0.03% of all ^{92}Kr beta decays are followed by the emission of a delayed neutron from the excited ^{92}Rb daughter. Because a short time interval (seconds to minutes) occurs between the beta particle emission and the neutron emission, it is called a delayed neutron. These delayed neutrons are useful in controlling nuclear reactions in power plants, as discussed further in the next chapter.

Other projectiles can induce fission reactions. This makes some sense, since the reasonable occurrence of fission requires only a big, excited nucleus. How that nucleus is formed is somewhat irrelevant. Protons, alpha particles, deuterons, and other small nuclei have caused fission in a variety of heavy targets. Since all of these projectiles carry a positive charge, they need to be accelerated before they can strike the heavy target (to overcome the Coulomb barrier). High-energy photons can also induce fission (**photo fission**) simply by transferring their energy to the nucleus. Neutron-induced fission is the most heavily studied because of its applications in nuclear reactors and weapons (Chapter 9), and because neutrons do not need to overcome the Coulomb barrier to react with a nucleus.

8.3   FUSION

Fission reactions release energy because a large nuclide splits into two that have greater stability in terms of binding energy per nucleon (Chapter 3). Likewise, the combination of two light nuclides forming a product with greater stability will also release energy. The combination of two light nuclides to form a heavier one is called **fusion**. Stars, like our Sun, are huge fusion reactors and the energy they emit comes from fusion reactions. Some of the important fusion reactions taking place inside stars and the total energy produced by each are given below.

$$_{1}^{1}\text{H} + _{1}^{1}\text{H} \rightarrow _{1}^{2}\text{H} + _{+1}e + \nu \qquad 0.42 \text{ MeV}$$

$$_{1}^{1}\text{H} + _{1}^{2}\text{H} \rightarrow _{2}^{3}\text{He} + \gamma \qquad 5.49 \text{ MeV}$$

$$_{2}^{3}\text{He} + _{2}^{3}\text{He} \rightarrow _{2}^{4}\text{He} + 2_{1}^{1}\text{H} \qquad 12.86 \text{ MeV}$$

It is so hot inside the Sun that matter does not exist as atoms—nuclei and electrons are separated and therefore it is plasma. It is more appropriate to think of the above reactions as taking place as nuclei, rather than atoms. The first reaction fuses two ^{1}H nuclei (protons) into a deuterium. Once formed, deuterium can combine with another proton to form ^{3}He (second reaction above). If the first two reactions run twice, we'll have two ^{3}He nuclei to fuse together to form ^{4}He and two protons.

The positrons produced in the proton–proton reaction above will annihilate (there are plenty of electrons in the Sun!), producing even more energy.

$$_{+1}e + _{-1}e \rightarrow 2\gamma \qquad 1.022 \text{ MeV}$$

Taken together, the four reactions shown here make up what is called the **proton cycle**. If we add them together, we will see the *net* reaction and energy for the proton cycle. Remember that the first two reactions need to happen twice, and therefore the positron annihilation also happens twice per cycle.

$$2(^1_1H + ^1_1H \rightarrow ^2_1H + _{+1}e + v \quad 0.42 \text{ MeV})$$

$$2(^1_1H + ^2_1H \rightarrow ^3_2He + \gamma \quad 5.49 \text{ MeV})$$

$$^3_2He + ^3_2He \rightarrow ^4_2He + 2^1_1H \quad 12.86 \text{ MeV}$$

$$2(_{+1}e + _{-1}e \rightarrow 2\gamma \quad 1.022 \text{ MeV})$$

$$2_{+1}^0e + 4^1_1H \rightarrow ^4_2He + 2v + 6\gamma \quad 26.7 \text{ MeV}$$

The proton cycle accounts for roughly 90% of the Sun's energy output. Fortunately our Sun is about 73% hydrogen (mass %), so it looks like there's still plenty of fuel up there. Notice that the Sun is also emitting a lot of gamma photons and neutrinos (Fig. 7.8).

We should also be thankful that the proton–proton reaction has a low cross-section (probability of success). This slows down the entire proton cycle. If the cross-section were higher, the Sun would burn hotter, and run out of hydrogen fuel faster—not an attractive prospect for life on Earth.

The fusion reactions shown above fit the general characteristics of the nuclear reactions studied in Chapter 7, but which is the projectile and which is the target? These reactions happen inside the Sun because the reacting nuclei are all moving very quickly, that is, they have a lot of energy. It is the high temperatures inside the Sun that allow these nuclei to overcome the Coulomb barrier and react. Because of this, fusion reactions are often called **thermonuclear reactions**.

The proton cycle is generally thought to be impractical for fusion power reactors here on Earth, primarily due to its relatively high **ignition** temperature—the temperature needed to start a fusion reaction. To get an idea of how hot it needs to be to get the proton–proton reaction going, let's calculate its effective Coulomb barrier.

$$^1_1H + ^1_1H \rightarrow ^2_1H$$

$$E_{ecb} \approx 1.11 \times \left(\frac{1+1}{1}\right) \times \left(\frac{1 \times 1}{\sqrt[3]{1} + \sqrt[3]{1}}\right) = 1.11 \text{ MeV}$$

This value can be converted to temperature using Boltzmann's constant:

$$1.11 \text{ MeV} \times \frac{10^6 \text{ eV}}{\text{MeV}} \times \frac{K}{8.63 \times 10^{-5} \text{ eV}} = 1.29 \times 10^{10} \text{ K}$$

According to these calculations, the Sun would have to be nearly 13 billion K in order to get the proton cycle to proceed in good yield. The surface of the Sun is about 6000 K, and the core is believed to only reach ~15 million K. Fortunately, plasma, like other forms of matter, contains individual particles that are moving both faster and slower than they should be at a particular temperature. Temperature is a reflection of the average kinetic energy of all particles, and there is always a distribution of energies about the mean value. Also remember that both particles on the left side of the arrow are moving, and if they hit head-on, they only need half as much energy each. Therefore, the Sun will have some particles moving fast enough to overcome the Coulomb barrier and fuse. Finally, remember that nuclear reactions can take place even if the particles lack the minimum energy—through tunneling. This has positive implications for human-built fusion reactors here on earth. These reactors do not need to get quite as hot as calculated using Boltzmann's constant, because some reactions will take place at lower temperatures due to tunneling.

The prospect of tunneling may have led two chemists to conclude they had observed fusion in 1989. They had observed heat being generated during an electrochemical experiment involving a palladium electrode in deuterated water (D_2O). They dubbed the phenomenon "cold fusion," then, circumventing the normal scientific peer review process, announced their results at a press conference. A great deal of popular media attention was given to this result, which seemed to promise an easy fix for our increasing energy needs. A number of other scientists tried to duplicate these remarkable results. At best, these efforts could be described as inconsistent, and at worst they were a clear repudiation of the original results. The scientists were discredited, and the entire scientific community suffered a loss of public confidence. Some research has continued in this area, but proof of low-energy nuclear fusion remains elusive. It is unfortunate that the term "cold fusion" was applied to these poorly interpreted experiments, as it is a bona fide area of nuclear science (see Section 8.5).

8.4 STELLAR NUCLEOSYNTHESIS

How were the elements we find here on Earth made? In the stars! Fusion reactions can make nuclides up to ^{56}Fe. Remember this is the pinnacle of nuclear stability (binding energy per nucleon—figure 3.1), so fusion up to the peak is exoergic, and endoergic after that.

After ^{56}Fe, stars rely on neutron capture to make heavier nuclides. The neutrons come from nuclear reactions taking place in the Sun like the one below.

$$^{21}Ne + {}^4He \rightarrow {}^{24}Mg + {}^1n$$

^{56}Fe will capture neutrons until it forms ^{59}Fe, which decays to ^{59}Co. ^{59}Co will capture a neutron and form ^{60}Co, which decays to ^{60}Ni. Through this combination of n, γ reactions and beta decay, almost all the naturally occurring nuclides from $A = 56$ through $A = 210$ can be prepared. This series of reactions is called the **s-process**, where s stands for slow. It's slow because the neutron flux inside of stars is not all that great. It takes a long time for significant amounts of nuclides beyond ^{56}Fe to build up.

The s-process also fails to explain the existence of a few stable, but somewhat proton rich, nuclides such as ^{124}Xe. ^{124}Xe cannot be formed through n,γ reactions followed by beta decay because it is "shielded" by stable ^{124}Te. Any neutron-rich, $A = 124$ nuclide will undergo beta decay until it forms ^{124}Te, which cannot then decay to ^{124}Xe. Nuclides such as ^{124}Xe are formed through the **p-process**. Two types of reactions are postulated to be part of the p-process: γ,n or p,x. Either one can lead to the formation of more proton-rich naturally occurring nuclides. Some of the other nuclides in the same boat are ^{78}Kr, ^{112}Sn, ^{120}Te, ^{144}Sm, and ^{184}Os. These nuclides are all found in lower percent abundances than the more neutron-rich stable isobars because the p-process is rather unlikely. This is generally true for two naturally occurring isobars between $A = 70$ and $A = 204$; the isobar with more neutrons will often have a higher percent abundance because it can be formed through the more likely s-process.

The s-process cannot make nuclides beyond $A = 210$. It terminates in the following cycle of reactions:

$$^{209}\text{Bi} + {}^{1}\text{n} \rightarrow {}^{210}\text{Bi} + \gamma$$

$$^{210}\text{Bi} \rightarrow {}^{210}\text{Po} + {}_{-1}e$$

$$^{210}\text{Po} \rightarrow {}^{206}\text{Pb} + \alpha$$

$$^{206}\text{Pb} + {}^{1}\text{n} \rightarrow {}^{207}\text{Pb} + \gamma$$

$$^{207}\text{Pb} + {}^{1}\text{n} \rightarrow {}^{208}\text{Pb} + \gamma$$

$$^{208}\text{Pb} + {}^{1}\text{n} \rightarrow {}^{209}\text{Pb} + \gamma$$

$$^{209}\text{Pb} \rightarrow {}^{209}\text{Bi} + {}_{-1}e$$

The rapid process (**r-process**) makes heavier nuclides. As its name implies, it is a series of quick neutron absorptions, leading to an extremely neutron-rich nuclide, which then undergoes a number of beta decays until the N/Z ratio optimizes. An extremely intense neutron flux is needed to accomplish this, as the neutrons need to be continually added, faster than the newly formed nuclides can decay. Such a flux is not possible under normal conditions in a star. It is believed that these fluxes are possible in supernovae. This could explain how heavy, naturally occurring elements such as Th and U are formed. Our solar system was formed from the remnants of a supernova, which explains why we have these heavy elements in some abundance. Therefore we live in a second (or possible third!)-generation solar system.

8.5   SYNTHESIS OF UNKNOWN ELEMENTS

Humans have done a pretty good job preparing isotopes of elements not found on Earth. These elements had been unknown because all of their isotopes have half-lives that are short when compared to the age of the solar system ($\sim$4.5 billion years). Therefore, if they were formed in the sun that preceded ours, they have since decayed away. The first "artificial" element in the periodic table is technetium (Tc, $Z=43$), and was first prepared in 1945 by accelerating a deuteron in a cyclotron and slamming it into a molybdenum target. Its name comes from the Greek word *technetos*, which means artificial. The preparation of technetium was a validation of the configuration of the periodic table as it filled a void in the middle of the transition metals. Technetium has since been made in large quantities as a fission product in nuclear reactors. A significant amount of its chemistry has been investigated, and it generally parallels the chemistry of its heavier congener rhenium. It is an obscure irony that technetium is now more plentiful on Earth than rhenium.

Promethium (Pm, $Z=61$) is the only other element below lead ($Z<82$) that was "missing" from the periodic table. It was first observed among the fission products of ^{235}U in 1945. It is named for the mythological Greek Titan who stole fire from the gods and gave it to humans.

Both Tc and Pm have an odd number of protons, which means that any of their isotopes with an odd number of neutrons ($A=$ even) will be unstable; but why are there no $A=$ odd ($N=$ even) stable isotopes of these elements? A simple way to look at Tc is to state that $^{98}_{43}Tc_{55}$ ($N/Z=1.28$) has the best N/Z ratio and the longest half-life of all the known Tc isotopes, but it is *oo*, and therefore cannot be stable. What about $^{97}_{43}Tc_{54}$ ($N/Z=1.26$) and $^{99}_{43}Tc_{56}$ ($N/Z=1.30$)? They have even numbers of neutrons and pretty good N/Z ratios. Examination of the $A=97$ and $A=99$ isobars shows that $^{97}_{42}Mo_{55}$ and $^{99}_{44}Ru_{55}$ are the only stable nuclides for those sets of isobars. As we saw in Chapter 4 (Section 4.6), when $A=$ odd there is only one stable isobar. Additionally, a stable ^{97}Tc or ^{99}Tc would violate Mattauch's Rule of no stable neighboring isobars.

What about the anthropogenic syntheses of the elements beyond uranium? A partial answer was given in an example in Chapter 7 where a plutonium target was hit with an alpha particle to produce a curium isotope.

$$^{240}_{94}Pu + {}^{4}_{2}He \rightarrow {}^{243}_{96}Cm + {}^{1}n$$

So long as the target nuclide can be isolated and has a reasonably long half-life, we can perform lots of reactions like the one above to make nuclides of the elements beyond uranium. We can even mimic the s-process, making new nuclides through n,γ reactions and beta decay, such as the preparation of ^{239}Pu from ^{238}U below.

$$^{238}_{92}U + {}^{1}n \rightarrow {}^{239}_{92}U + \gamma$$

$$^{239}_{92}U \rightarrow {}^{239}_{93}Np + {}_{-1}e$$

$$^{239}_{93}Np \rightarrow {}^{239}_{94}Pu + {}_{-1}e$$

The r-process is mimicked in the explosions of thermonuclear weapons. As we'll see in the next chapter, tremendous neutron fluxes are briefly created in these explosions that can add several neutrons to ^{238}U, which is often used as a casing material for these weapons. The first detonation of a thermonuclear device occurred in 1952 on Eniwetok Island in the South Pacific. Two elements that were first observed in the debris following this explosion were einsteinium (Es) and fermium (Fm). As many as 17 neutrons were absorbed by ^{238}U atoms in the bomb's casing during the explosion, followed by a series of beta decays to form the observed nuclides.

We can only get so far with the preceding reactions. Generally speaking, the half-lives of nuclides get shorter as atomic number increases beyond uranium. This makes intuitive sense—the nuclides are already too big, and they are getting bigger. Larger nuclides mean the strong force is having an increasingly difficult time keeping the nucleus together under the burgeoning Coulomb repulsion of all the protons. At a certain point, the half-lives of potential targets will get too short, so the projectile will have to be bigger if we want to make super-heavy nuclides. For example, ^{254}No can be prepared by hitting ^{246}Cm with a ^{12}C projectile:

$$^{246}_{96}\text{Cm} + ^{12}_{6}\text{C} \rightarrow ^{254}_{102}\text{Pu} + 4\,^{1}\text{n}$$

Increasing the Z of the projectile increases the Coulomb barrier and the excitation energy of the compound nucleus. We'll soon reach a point where the compound nucleus will have too much excitation energy and will fall apart in ways that make it difficult to determine that it was actually made. A more subtle approach is required.

If the projectile has less than the minimum energy required for the product to be formed in good yield, the product can still be formed, there just won't be very much of it. Thankfully, nuclear decay can be detected, even if only a handful of decays occur—so we really only need to be able to make a few product nuclides. This approach has traditionally been referred to as "cold fusion," while the ^{246}Cm example above would be called hot fusion. As mentioned earlier in this chapter, cold fusion has unfortunately become more widely applied to alleged low-energy nuclear reactions involving low-Z nuclides. The first atoms of element 111 (roentgenium, Rg) were made with the cold fusion method, bombarding a ^{209}Bi target with ^{64}Ni:

$$^{209}_{83}\text{Bi} + ^{64}_{28}\text{Ni} \rightarrow ^{272}_{111}\text{Rg} + ^{1}\text{n}$$

QUESTIONS

1. Calculate the partial half-lives for the two modes of decay (alpha emission and spontaneous fission) for ^{238}U. How many alpha and spontaneous fission decays take place every minute inside a 100 g block of ^{238}U?

2. $^{256}_{100}$Fm can decay by spontaneous fission, yielding two atoms of $^{128}_{50}$Sn. Calculate the energy released in this reaction and estimate the kinetic energy of the tin fragments. Why is symmetric fission of ^{256}Fm a reasonable possibility?

3. $^{240}_{94}$Pu can decay by spontaneous fission yielding ^{103}Mo and ^{134}Te. Write a balanced equation and estimate the kinetic energy of the fragments. Would you expect the fission yield for these products to be high? Why?

4. Write out the decay series that is generated by the formation of ^{140}Cs as a fission product. Calculate the N/Z ratio for each isobar formed.

5. Calculate the excitation energy of the compound nucleus formed by the capture of a thermal neutron by ^{239}Pu and ^{240}Pu. If the fission barrier for ^{240}Pu is 5.5 MeV, decide which (if either) nuclide will fission. If neither nuclide will fission, how much energy would need to be brought in by the neutron to overcome the fission barrier?

6. Calculate the minimum neutron energy needed for fission of ^{238}U to proceed in good yield. Don't forget to account for the kinetic energy of the compound nucleus.

7. One of the possible reactions for thermal neutron fission of ^{239}Pu produces ^{134}Te and ^{103}Mo. Write out a balanced equation, and calculate the total energy released and estimate the kinetic energies of the products.

8. Using the reaction in the previous problem, calculate the nuclear binding energy for ^{239}Pu and the products. Are the differences between reactants and products consistent with the results above? Explain.

9. Ten years after some pure ^{235}U undergoes fission, what are the major radioactive nuclides present?

10. Home fusion enthusiasts claim they can perform the reaction below in their garage by accelerating deuterons through a 45 kV electric field. Estimate the effective Coulomb barrier for this reaction and comment on the validity of their claims. If the reaction below is taking place in someone's garage, what safety concerns might result? Also calculate the energy produced by this reaction.

$$2\,{}^{2}_{1}\text{H} \rightarrow {}^{3}_{2}\text{He} + {}^{1}\text{n}$$

11. In 2002, a scientist at Oak Ridge National Laboratory claimed he was observing the fusion reaction below by exposing deuterated acetone $((CD_3)_2CO)$ to 14 MeV neutrons and ultrasound. Suggest another reaction that could've produced the tritium.

$$2\,{}^{2}_{1}\text{H} \rightarrow {}^{3}_{1}\text{H} + {}^{1}_{1}\text{H}$$

12. Starting with ^{131}Xe, give the likely nuclear reactions that might produce ^{132}Xe and ^{132}Ba in stars. Look up the percent abundances of ^{132}Xe and ^{132}Ba. Can you explain the difference between them?

13. Rationalize why no stable isotopes of Pm exist.

14. ^{244}Pu was detected for the first time following the thermonuclear detonation on Eniwetok Island. Show the probable nuclear reactions and decays that likely led to the formation of this nuclide from ^{238}U. Also show how ^{235}U might be formed in a supernova from ^{209}Bi. This time simplify the process by writing it as a single nuclear reaction.

15. ^{266}Mt can be prepared by the reaction of ^{209}Bi with ^{58}Fe. Write out the complete nuclear reaction for this synthesis. What projectile energy would be required to prepare ^{266}Mt in good yield using this reaction? What would the excitation energy of the compound nucleus be in this case? Do you think this reaction is an example of hot or cold fusion? Briefly explain. If the excitation energy of the compound nucleus were limited to 13 MeV, what would the projectile energy be?

9 Nuclear Reactors and Weapons

Nuclear fission and fusion are special types of nuclear reactions because of their use, or potential use, in power generation and in weapons of mass destruction. Nuclear reactors and weapons continue to be controversial issues in the United States, and are often mischaracterized by those taking strong positions concerning their use. This chapter seeks to explain some of the science and thereby deflate some of the popular political posturing surrounding these hot button topics in an impartial manner, in hopes that those who read through it might begin to appreciate their complexity.

9.1 FISSION REACTORS

As we saw in the previous chapter, a tremendous amount of energy is released ($\sim$200 MeV) in every fission reaction. Nuclear power plants convert some of this energy to electricity to help satisfy the world's increasing demands for more power. The key to nuclear power production is the fact that it can be a self-sustaining **chain reaction** (fig. 9.1). On average, 2.4 neutrons are produced by the fission of each ^{235}U nuclide. If these neutrons can find additional ^{235}U nuclei to run into, they could cause more fission reactions to take place, producing still more neutrons, to cause still more fission reactions. Not all neutrons produced will cause additional fission reactions—some will escape the fissile material, and others may cause other reactions (such as n, γ). In figure 9.1, notice that we start with just one neutron at the top, which produces two neutrons, then these two neutrons produce four neutrons, which produce eight neutrons, and so on. The number of fission reactions that take place will rapidly increase with each subsequent generation.

The key is to have enough ^{235}U nearby to make the reaction self-sustaining. The minimum amount of material to obtain a self-sustaining chain reaction is known as the **critical mass**. For pure nuclides of the same shape, this value will depend on the average number of neutrons produced per fission, and the probability of fission (σ_f). ^{239}Pu produces an average of 2.9 neutrons per fission (2.4 for ^{235}U) and has a fission cross-section value of 752 b ($\sigma_f = 586$ b for ^{235}U), and therefore has a smaller critical mass than ^{235}U. If less than the critical mass is present, it is called a subcritical mass, and cannot sustain a chain reaction. If more than the critical mass is present, it is a **supercritical mass**. Uncontrolled, a supercritical mass can generate enough energy for a very large explosion.

A nuclear power plant contains a mass of material that, if it was all placed together, would constitute a supercritical mass. The material is separated in such a way that it can achieve **criticality**—a self-sustaining chain reaction—without an exponential increase in fission reactions (**supercriticality**). Cylindrical pellets, about the size of the end of your little finger, are placed end to end in a metal pipe that is

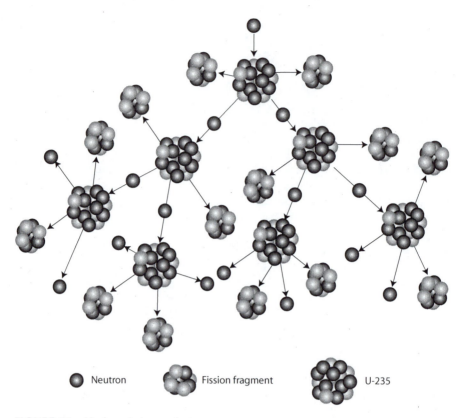

Neutron   Fission fragment   U-235

FIGURE 9.1   Nuclear chain reaction.

then sealed at both ends (**fuel rod**), and these rods are then distributed in the reactor parallel to each other and separated so there's space between them (fig. 9.2).

The goal of the power plant is to maintain the rate of fission reactions at a constant level. If there are too many neutrons flying around, the reaction will generate too much thermal energy. If the reactor gets too hot, the fuel rods (and the metal assemblies that hold them) will melt (**reactor meltdown**). If there are too few neutrons, the chain reaction will shut down. To maintain the perfect number of neutrons, some will have to be absorbed. This is done with **control rods**, which can move in and out from between the fuel rods (fig. 9.2). The control rods are typically made out of materials that have a very high neutron absorption cross section (σ_a), such as cadmium ($\sigma_a = 2520$ b) or boron ($\sigma_a = 760$ b). The more control rods that are lowered further down, the slower the overall rate of fission reactions. The more the control rods are raised, the greater the number of fission reactions that occur per unit time.

As mentioned in Chapter 8, not all neutrons produced in a nuclear reactor originate in the fission reaction. Some of the fission products emit delayed neutrons a short time (seconds to minutes) after the fission occurs. These neutrons can also cause more fuel to undergo fission, and are important in controlling the rate of fission reactions in the reactor. If fission were the only source of neutrons (**prompt neutrons**), the reactor operators would only have a fraction of a second to respond if the number of fission reactions suddenly started to increase. Even computer-operated control rods

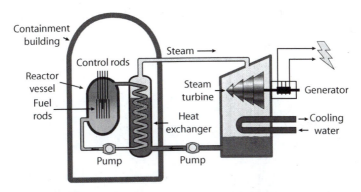

FIGURE 9.2   Schematic drawing of a nuclear power reactor (PWR).

would not have enough time to respond, and meltdowns would be common. The fact that a small fraction of all the neutrons flying around the reactor are produced some time after fission occurs allows operators and automated safety systems a minute or two to respond to an escalation in the chain reaction. The delayed neutrons provide a nice buffer to sudden changes in the number of fission reactions per unit time.

As you remember from Chapter 7, higher-energy neutrons generally have a lower probability of inducing a reaction than slower neutrons. In Chapter 8, we learned that the average energy of a neutron produced in the fission of ^{235}U is 2 MeV. If more fission reactions were desirable, it'd be best to slow them down. This is another good reason to separate the fuel rods. Aside from being able to place the control rods between the fuel rods, it would also be a good idea to put some stuff in there that will slow down the neutrons. Since this material is slowing the neutrons, it is called a **moderator**. An ideal moderator has a relatively low neutron absorption cross section (σ_a), yet has a high neutron scattering cross section (σ_s), and is relatively inexpensive. Water (H_2O) is commonly used as a moderating material as its σ_a is 0.66 b, its σ_s is 49 b, and it is very cost-effective. Other moderator materials are D_2O ($D = {}^2H$, $\sigma_a = 0.0013$ b, $\sigma_s = 10.6$ b) and graphite (a form of pure carbon, $\sigma_a = 0.0035$ b, $\sigma_s = 4.7$ b). Based solely on their very low neutron absorption numbers, both D_2O and graphite would appear to be superior moderators to H_2O, but water is less expensive, and it can also be used as a **coolant** for the reactor. The coolant carries the thermal energy away from the fuel to be used in power generation. A loss of coolant can cause enough thermal energy to build up to cause a reactor meltdown. The biggest concern of a meltdown is that it may cause a breach in the containment structure, releasing radioactive materials into the environment (see Section 9.2).

When compared to a fossil fuel power plant, nuclear plants look very similar. Fossil fuel plants burn hydrocarbon (chemical compounds containing both hydrogen and carbon) fuel. The energy released from the chemical reaction (combustion) that takes place heats water to steam, which drives a steam turbine that produces electricity. Nuclear reactors use nuclear reactions (fission) to do the same thing (fig. 9.2). The main difference between fossil fuel and nuclear plants is how the water is heated.

Careful examination of figure 9.2 will show that the water heated by the nuclear reactions is not the same water (steam) that drives the turbines. The water heated by the nuclear reactions is kept under pressure, so that it remains a liquid. It is pumped through the reactor **core** (where the fuel is), then through a heat exchanger to transfer its thermal energy to another closed water loop. This second loop is operated at lower pressure, so the water is heated to steam, which then drives the turbines to produce electricity. This water then needs to be cooled (condensed back to liquid), so it runs through another heat exchanger, transferring its excess thermal energy to a third loop of water. Unlike the first two loops, the third loop is not closed. It transfers thermal energy to the environment—a river, lake, ocean, or the air. The immense, hour-glass shaped cooling towers that are so often associated with nuclear power plants simply transfer this waste heat to the air. The white clouds that are often seen emerging from them are not smoke, but steam.

The reactor depicted in figure 9.2 is called a pressurized water reactor (PWR), because water is used as a coolant and it is kept under pressure in the reactor core. Roughly two-thirds of the world's nuclear reactors have a PWR design. Another popular design (most of the remaining reactors) is the boiling water reactor (BWR—fig. 9.3). The boiling water reactor doesn't pressurize the water that is pumped into the core, so it boils. The steam that is generated drives the turbines directly instead of transferring its thermal energy to a second water loop. The steam still needs to be condensed and the waste heat transferred to the environment.

Dealing with waste heat is also an issue for fossil fuel power plants. Coal or natural gas plants also need to be located near large bodies of water to soak up the waste heat, or have cooling towers to dump the waste heat into the air. Dumping waste heat into a body of water can have deleterious environmental consequences. To avoid these problems, the water from the third loop can be pumped into an artificial holding pond before releasing it to natural waters.

As the previous paragraph implies, not all of the energy released by the fission reactions can be converted to electricity, that is, the process is not 100% efficient. In fact only ~35% of the energy released by the nuclear reactions is successfully converted to electrical energy. This is comparable to efficiencies obtained in most fossil fuel power plants.

Where does nuclear fuel come from? Rocks, mostly. Naturally occurring uranium ores are common—the largest proven reserves are currently in Canada, Australia, and Kazakhstan. The uranium is isolated via chemical reactions and separations from some of the other elements present in its ores. It is isolated as U_3O_8, which is

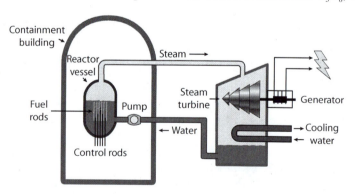

FIGURE 9.3   Schematic for a boiling water reactor (BWR).

also known as "yellow cake," probably because it is yellow powder. Naturally occurring uranium is only 0.72% ^{235}U; almost all of the rest (99.27%) is ^{238}U, which does not undergo fission with thermal neutrons.

Nuclear reactors need the uranium to be ~4% ^{235}U. Increasing the percentage of one isotope relative to others is called **enrichment**, and typically requires a physical (rather than chemical) process. For many years, the process of choice for uranium enrichment was **gaseous diffusion**. The U_3O_8 is chemically transformed to uranium hexafluoride (UF_6). UF_6 is a solid under normal conditions (1 atm, 20°C), but sublimes to a gas if the temperature is raised just a bit or the pressure is lowered. Because the $^{235}UF_6$ molecules are a little lighter than the $^{238}UF_6$ molecules, they move (diffuse) a little faster. If we pass UF_6 gas through a very long pipe, the gas that emerges at the far end will have a higher percentage of $^{235}UF_6$ relative to $^{238}UF_6$. The pipe has to be pretty darned long, as the difference in mass is pretty subtle. The relative diffusion rates can be calculated using Graham's Law.

$$\frac{rate_1}{rate_2} = \sqrt{\frac{molar\ mass_2}{molar\ mass_1}} \tag{9.1}$$

$$\frac{rate_{^{235}UF_6}}{rate_{^{238}UF_6}} = \sqrt{\frac{MW_{^{238}UF_6}}{MW_{^{235}UF_6}}} = \sqrt{\frac{352.04\ g/mol}{349.03\ g/mol}} = 1.0043$$

$^{235}UF_6$ only diffuses 1.0043 (or 0.43%) faster than $^{238}UF_6$! Extremely large facilities were constructed to accommodate the tremendous amount of plumbing required. The first such plant built in the United States was the K-25 facility near Oak Ridge, Tennessee. It was four stories tall, half a mile long, and a thousand feet wide. A huge building!

Today, enrichment using gas **centrifuges** is more common. UF_6 is placed in a cylinder that is spun around at high speed. The heavier $^{238}UF_6$ tends to collect near the cylinder wall, while the lighter $^{235}UF_6$ tends to collect near the central axis of the cylinder (fig. 9.4). Gas centrifuges are preferred for uranium enrichment because they require only about 3% of the power when compared to an equivalent gaseous diffusion plant. Gas centrifuge plants are also much smaller than gaseous diffusion plants, and are therefore attractive to those who wish to be somewhat secretive about their enrichment activities.

Once enriched, the uranium is converted back to an oxide (UO_3) and compressed into small pellets for nuclear fuel. If the uranium is to be used in a weapon, it needs to be >80% ^{235}U, which means more processing, followed by a chemical transformation to uranium metal.

The uranium left over from the enrichment process is ~99.7% ^{238}U, and is referred to as **depleted uranium** (or DU). Because of its exceptionally high density (1.7 × greater than lead!) and its relatively low radiological danger, depleted uranium is sometimes used as ballast in boats, counterweights, or, ironically, radiation shielding. When combined with nitrogen under the right conditions, depleted uranium forms an exceptionally hard and dense material (uranium nitride). Uranium nitride

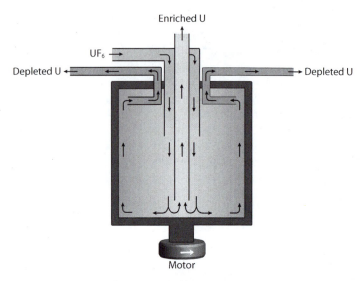

FIGURE 9.4   Flow schematic for a gas centrifuge.

is used as a coating for armor piercing shells. These shells were used extensively by the United States in its Gulf Wars, which may result in a significant health hazard to those exposed to uranium-containing vapors or shrapnel when they were used to destroy Iraqi tanks. There is also a concern that shrapnel left in the desert may represent an environmental hazard.

Can other fuels (besides $\sim 4\%$ ^{235}U in ^{238}U) be used in nuclear reactors? Certainly! Any fissile material will work, with ^{235}U, ^{233}U, and ^{239}Pu being the most popular. ^{239}Pu is both produced and "burned" (undergoes fission) in most nuclear power reactors. As outlined at the end of Chapter 8, it is produced through neutron capture and subsequent decay of ^{238}U:

$$^{238}U + n \rightarrow {}^{239}U + \gamma$$

$$^{239}U \rightarrow {}^{239}Np + {}_{-1}e$$

$$^{239}Np \rightarrow {}^{239}Pu + {}_{-1}e$$

Reactors can be designed for ^{239}Pu production, or it can simply be separated from the reactor fuel pellets. Highly enriched uranium or ^{239}Pu are generally used for nuclear weapons (see Section 9.7); however, obtaining ^{239}Pu from reactors is much easier than enriching uranium to $>80\%$, which is necessary for weapons. Instead of a tedious physical process, plutonium can easily be separated from uranium via simple chemical processes. Its ease of isolation has made ^{239}Pu the nuclear weapons material of choice and the proliferation of nuclear weapons all too easy.

The dramatic reduction of the U.S. and Soviet nuclear weapons arsenals following the end of the Cold War has generated a great deal of highly enriched uranium and ^{239}Pu. Much of the nuclear materials from the decommissioned weapons are being securely stored at the Y-12 plant in Oak Ridge, Tennessee. These materials can be used for reactor fuel. For security purposes, the uranium will be blended with depleted uranium to bring the ^{235}U down to $\sim 4\%$, and the ^{239}Pu can be mixed with reactor-grade uranium to produce a mixed oxide (MOX) fuel. While MOX fuel is new to the United States, Europeans have been burning MOX fuel in their reactors for many years.

A high-profile program called "Megatons to Megawatts" was initiated between Russia and the United States in 1993. Under this program, the United States purchases highly enriched uranium from Russian warheads and converts it to reactor fuel for use in U.S. reactors. As of this writing, roughly half of the fuel in U.S. nuclear reactors came from Russian weapons that used to target the United States. This means that roughly 10% of all electricity in the United States currently originates from a former Russian weapon. A wonderful irony!

Approximately 20% of the U.S. electrical production is nuclear, while in some European countries it accounts for the vast majority of all power generation. There has been no licensing of new nuclear power plants in the United States since the mid-1970s. This was a result of the relatively high cost of nuclear power, the accidents at Three Mile Island and Chernobyl, and the federal government's lack of resolve in coming up with a plan to deal with spent fuel. As a result, these plants tend to be rather old. Because they have a limited lifetime, some are shutdown, while others have sought license renewals to continue beyond their originally intended period of operation.

The U.S. government began to quietly encourage the licensing and construction of new plants in the late 1990s and early 2000s. This change in attitude results from increasing costs of oil and natural gas, which currently account for $\sim 30\%$ of all electricity production in the United States, and increasing concerns that burning fossil fuels is having a deleterious effect on the global environment. Ironically, nuclear power, which has long been opposed for environmental concerns (potential for a meltdown, dangers from waste), is now being promoted as a relatively green power source because it does not produce greenhouse gases. Regardless of how it is viewed, it seems that new nuclear power plant construction in the United States now appears inevitable. As of this writing, about 30 plants are being considered for construction (mostly in the Southern United States), but no construction has begun. The following sections of this chapter discuss the four significant issues surrounding nuclear power: safety, waste, cost, and proliferation of nuclear weapons.

9.2 REACTOR SAFETY

While their safety record is quite good when compared to fossil fuel plants, the American public generally views nuclear power plants as dangerous. A small portion of this concern is justified, but almost all of it is based on fear and ignorance. Since concerns related to waste and proliferation are discussed below, only issues related to the mining of uranium and the potential of a catastrophic accident are mentioned here.

Like other forms of mining, uranium mining is hazardous. Because uranium miners now take preventative measures, and their radiation exposure is carefully prevented and monitored, they face little additional risk when compared to other miners. Unfortunately this was not true in the early days of uranium mining. In the 1950s, miners were exposed to and ingested doses that would be unthinkable today. Decades later these uranium miners now have cancer rates that are significantly higher than the national average.

Currently, there is very little uranium mining going on in the United States. Most of the proven deposits were mined out by the 1990s. The biggest current risk from uranium mining in the United States has to do with mine tailings. Tailings are the solid material left over at the mine after the uranium has largely been removed. In some cases these tailings have been left on the surface, and can leach radioactive nuclides into local water supplies. These sites are now being cleaned up, but hundreds of people have been exposed, and for a small fraction this exposure will result in a premature death.

Chernobyl is the only major nuclear power plant accident known to cause significant human fatalities. Despite the magnitude of this accident, and the fear it has engendered, only about 30 deaths immediately followed the accident. These deaths were among the emergency response personnel that received a very large dose while trying to contain the disaster. A couple of hundred to a few thousand premature deaths are also expected—primarily due to thyroid conditions in those who were children living near the reactor at the time of the accident.

So, yes, nuclear power has proven hazardous, and may continue to do so for future millennia if our solution to the waste disposal issue proves inadequate (vide infra). How does nuclear power stack up against other forms of electrical generation? Tens of thousands of people die each year due to coal's use as the most significant form of power generation in the world today. Coal mining remains one of the most dangerous occupations, and the transportation and burning of the large amounts of coal necessary to keep the plants running remains dangerous for the general public. The burning of coal and other fossil fuels adds to the dramatically increasing concentration of CO_2 in Earth's atmosphere, which is clearly contributing to global climate change. If just some of the recent natural disasters (such as Hurricane Katrina) were aggravated by global warming, then the safety record of fossil fuels is even worse than stated at the beginning of this paragraph. Coal also contains some uranium and thorium (naturally)—if not trapped, they can come out of the stack to be distributed downwind. It is a sad irony that people living downwind of a coal-burning plant can receive a greater radiation dose than those living near a nuclear power plant. If fossil fuels were held to the same standards of safety as nuclear power, almost all of our power today would come from nuclear.

Hundreds of nuclear plants around the world have been operating for decades with excellent safety records. Even so, it might be helpful to more closely examine the two most significant nuclear power plant accidents—Three Mile Island (TMI) in 1979 and Chernobyl in 1986. Both accidents were the result of a loss of water coolant, which caused melting of the fuel rods (only partial for TMI). The TMI reactor core material did not breach the containment building, however, $\sim 10^{12}$ Bq of radioactive gases (mostly Xe and Kr) were vented during the crisis. Because it was all in the gas phase, it dispersed quickly, resulting in minimal doses to people living nearby. Radiation levels outside of the plant were carefully monitored, and at no time was there a significant danger to the public. As a result, mandatory evacuation orders were never issued, which was fortunate, as the resulting panic would've undoubtedly resulted in injury and loss of life.

It is a tragic irony that it was once thought that a loss of water coolant could not lead to a meltdown. Since water also moderates the neutrons, it was thought that the fission chain reaction would shut down on its own because of the lower probability of fast neutrons inducing a fission reaction. TMI clearly demonstrates this is not the case. This has reduced the credibility of those who now claim that new reactor designs cannot meltdown.

The Chernobyl accident was much worse, primarily because the reactor used graphite, instead of water, as a moderator. It also suffered from a known flaw in its design that allowed for rapid increases in power output under certain conditions. Unfortunately, these conditions occurred at a time when reactor operators had turned off the coolant (water) pumps to conduct an experiment. The rapid rise in core temperature resulted in two steam and/or hydrogen explosions that blew the roof off the containment building, and then the graphite moderator caught fire. Eventually the fuel melted through the floor of the containment building, but it was the persistent graphite fire that led to the extensive distribution of radioactive materials. An estimated 2×10^{18} Bq of radioactivity was released. Much of it was dropped on Northern, Eastern, and Central Europe, although measurable quantities were spread throughout much of the Northern Hemisphere.

The operation and design of current and future nuclear plants are imminently conscious of mistakes and design flaws of the TMI and Chernobyl reactors, and have attempted to make reactors safer. A considerable amount of time has passed since the Chernobyl (1986) disaster without another accident resulting in public exposure, suggesting that new designs and procedures are working.

9.3   NUCLEAR WASTE

When fuel rods are first removed from the reactor core, they are still pretty toasty—very radioactive and still producing a fair amount of thermal energy (hot!), both of which are due primarily to the presence of short-lived fission products. They are stored under water at the reactor site while the short-half-life nuclides decay and the rods cool off. The water acts as shielding for the radiation as well as a coolant. After they've cooled off a bit, the rods can be placed in dry cask storage. These casks are designed to allow air-cooling (convection) of the rods, provide shielding, and be generally indestructible.

What is spent fuel composed of? Remember that the fuel starts out as $\sim 4\%$ ^{235}U in ^{238}U. When the ^{235}U burns in a nuclear reactor, the percentage of ^{235}U gradually decreases. When fuel rods can no longer be used (**spent**), they still contain roughly 1% ^{235}U and 1% ^{239}Pu, which could still be burned in a reactor. Only $\sim 3\%$ of the spent fuel are fission products, and the remaining 95% is still ^{238}U. Since the fission products generally have shorter half-lives than the uranium and plutonium, and the ^{235}U and ^{239}Pu can still be used as fuel, some have suggested that separations be performed on the spent fuel. This is known as **reprocessing**, and is the approach some countries have taken toward spent fuel. It should be noted that reprocessing is much more expensive than mining uranium and enriching it to new fuel and, as discussed below, presents a proliferation problem. The major motivations for reprocessing are environmental. Reprocessing will significantly reduce the volume of waste needing to be stored and allow more of the waste to become cold (no longer radioactive) sooner.

Since the 1970s, the United States had intended to simply dispose of its spent fuel as radioactive waste. This is called the "once-through" fuel cycle because the fuel is only in the reactor once. The U.S. federal government promised companies running nuclear reactors that it would take all of the spent fuel and bury it in a geologic repository. For many reasons (some discussed below) the U.S. government was unable to meet its own deadlines to open the repository, and all spent fuel currently remains on site at the reactors. In a move that may in part be designed to make opening the repository more politically palatable, the U.S. government reversed its 30-year policy on spent fuel in 2006, deciding in favor of reprocessing.

Even with reprocessing, there will still be a large volume of radioactive waste to dispose of, and geologic repositories (burying it) are always the proposed solution. Why? Many options have been studied, from dumping near a subduction zone, to launching it into the sun. Geologic repositories have been determined to be the safest way to dispose of nuclear waste.

The best support for a geologic repository originally came from a uranium mine. The French get some of their uranium from the African nation of Gabon. Back in the 1970s, a shipment from the Oklo mine in southeastern Gabon arrived at the French enrichment plant with ~0.5% ^{235}U (should be 0.72%). Other shipments showed higher than normal percentages of ^{235}U. French scientists started looking into this and found that other isotopic ratios at the Oklo site were not what they should be. Through a great deal of analysis, they reasoned that about 1.7 billion years ago, a fission chain reaction took place. In other words, Oklo was the site of a natural nuclear reactor. All the conditions were right. At that time, the ^{235}U natural abundance was ~3%, and there was plenty of water around to moderate the neutrons. It looks like the Oklo reactor ran for about 100,000 years (off and on, depending on the presence of water), burning about 6 tons of ^{235}U and generating about 1 ton of ^{239}Pu (which underwent fission, or decayed to ^{235}U) and 15,000 MWy of power—roughly equivalent to four years' output of a modern reactor. It also appears that almost all of the radioactive nuclides produced by the Oklo reactor stayed right where they were made, or traveled only a short distance, suggesting that geologic repositories may be an effective long-term storage solution for radioactive wastes.

The first such repository opened in 1999 near Carlsbad, New Mexico. It is called the "Waste Isolation Pilot Plant" (WIPP), and is constructed in a 200 million-year-old salt formation ~2000 ft underground. WIPP cannot accept spent nuclear fuel, but does accept less radioactive (low-level) wastes from research and other activities involving radioactive materials. As suggested above, a good repository is geologically stable and dry. If WIPP had any water flowing through it, the salt would've dissolved. Another consideration is security. Having all this waste in one isolated place is safer than several more accessible places. Finally, politics are also a consideration. New Mexico is sparsely populated and does not have a considerable amount of political clout in Washington DC, making it easier for the federal government to establish a waste site there. Additionally, much of the waste being sent to WIPP has been generated at Los Alamos National Laboratory, also located in New Mexico.

WIPP is unsuitable for spent fuel because the heat from waste will draw water to it, bathing the containers in salt water. Under these conditions, they would quickly corrode, and potentially release their contents to the environment. Yucca Mountain, near Las Vegas in Nevada, has been identified as a potential site for a geologic repository for spent fuel and other more radioactive (high-level) wastes. Yucca Mountain shares many of WIPP's positive attributes except that it is not a salt formation. Additionally, Yucca Mountain is part of the Nevada Test Site, where nearly 1000 nuclear weapons were tested both above and below ground for over 40 years—until the test ban in 1992. This site is large, isolated, and secure. Opening of Yucca Mountain has been delayed, and it now looks unlikely to open before 2021. One of the major causes of the delay has been concern that the geology might not be as stable or dry as originally thought, especially over the long-term (millions of years!). Reprocessing the spent fuel to remove the longer-lived radioactive nuclides may help alleviate this concern. The other cause for delay has been Nevada's increasing political power, which has often inhibited key legislation that would facilitate opening of the site.

Despite the fact that there is a lot of it (thousands of tons) and that it is rather toasty, spent fuel is well characterized, rather compact, and quite uniform (it's the same stuff no matter what reactor it comes from). The other major source of high-level waste in the United States has none of these positive attributes. It is a legacy of the Cold War with the Soviet Union—making thousands of nuclear weapons generates a lot of radioactive trash, as well as contaminated buildings and grounds. By far, the biggest challenge is cleaning up the 177 underground waste tanks at the Hanford site in south central Washington State. Hanford was started as part of the Manhattan Project and provided the plutonium for the first test weapon as well as the bomb dropped on Nagasaki. During the Cold War, the Hanford site was responsible for producing the plutonium and tritium (^{3}H) for the U.S. nuclear arsenal. Eight nuclear reactors were built along the banks of the Columbia River and plutonium separation facilities were placed further inland.

Separation of plutonium is accomplished through a chemical process. At the end of this process, nitric acid solutions containing fission products, as well as traces of uranium, plutonium, other actinides, and lanthanides needed safe disposal. The tanks were built to accommodate these wastes. The tanks were made of steel, and would not hold acidic solutions (low pH) for long, so enough sodium hydroxide (NaOH, aka lye) was added to make the waste solution basic (high pH), and therefore not corrosive to the tanks. Tank volume occasionally became scarce, so the water (not radioactive) was evaporated out of many, so additional waste could be added. So much water was evaporated that the tanks now all contain a lot of solids suspended in aqueous slurries (sludge) or hardened into salt formations (saltcake). Radioactive materials are part of or entrapped in both.

Different waste streams were added to different tanks, and sometimes the contents of one tank were pumped into another. Records of these activities are incomplete, leaving each tank with a unique and partially unknown chemical and radionuclide composition. One tank contains a fair amount of organic compounds (containing mostly carbon and hydrogen). The radiation causes the formation of hydrogen gas (H_2) from the organic compounds. The H_2 would build up in the sludge, and then occasionally "burp" out of the tank. Concerns were raised that the hydrogen could cause the tank to explode during a burp, so a method of stirring the tank contents was developed to allow the slow continuous release of the hydrogen gas.

Most of the materials in these tanks are not radioactive. The three most common ions present are sodium (Na^+), hydroxide (OH^-), and nitrate (NO_3^-). Various methods of separating the hot (radioactive) from the cold (not radioactive) have been researched and some implemented. The idea is to place the cold stuff into a relatively inexpensive waste form, such as concrete (or grout), on site and **vitrify** the relatively small volume of radioactive material. Vitrification means isolating the radioactive stuff in glass, and is a relatively expensive process. The radioactive material is distributed throughout a large glass cylinder (called **logs**) on site then shipped to a high-level repository such as Yucca Mountain. Glass is the preferred waste form for long-term storage of radioactive nuclides, as even if water flows over it, it is unlikely any of the radioactive material will leach out.

The engineers that designed the tanks thought they would only be used temporarily, not the 60+ years some have been in use. Some of the tanks have cracked and leaked some of their contents into the ground nearby, creating a more difficult clean-up scenario.

Similar waste tanks also existed at Department of Energy (DOE) facilities near Idaho Falls, Idaho, Oak Ridge, Tennessee, Aiken, South Carolina, and West Valley, New York. These sites only contained a few tanks each; most have already been remediated or will soon be cleaned up.

Hospitals, universities, and private companies produce small amounts of very low-level radioactive wastes. These wastes can often be disposed of in local landfills, or landfills specially designed for the disposal of hazardous material, that is, extra effort is taken to isolate the contents of the landfill from the surrounding environment.

9.4   COST OF NUCLEAR POWER

When nuclear power plants were first being planned in the United States, the statement that "nuclear power would be too cheap to meter" was made. The fact that it subsequently became one of the more expensive ways to generate electricity damaged the credibility of the industry, and is probably the single biggest factor in the lack of construction of new plants in the United States since the mid-1970s. With increases in costs of oil and natural gas, coupled with increasing concerns that the extensive burning of fossil fuels is adversely affecting the global climate, interest in nuclear power has grown in the United States.

The average U.S. wholesale cost of generating electricity from all sources in 2005 was $50/MWh (megawatt hour—the production of one million watts of power for one hour). The same costs for U.S. nuclear power plants varied from $30/MWh to $140/MWh in 2005. Why such a huge spread? The single most important factor in the cost of nuclear power is the time it takes to build the plant. Nuclear power is very capital intensive—a lot of money is required up front. Once it is up and running, nuclear plants are relatively cheap to operate because their fuel costs are low compared to fossil fuels—a lot of energy is generated from very little fuel. Time is an issue because most of the capital costs are covered through loans. If the plant takes a long time to build, the interest on the loans can become as large as the loan itself. These costs are recovered only after the plant is built, and are passed on to the ratepayers. This was especially bad through the periods of high inflation like the 1970s and 1980s. The worst example is a plant at Watts Bar in Tennessee. It was one of the last licenses granted (1973) in the United States and was the most recent plant completed (1996!).

It is estimated that new nuclear plants constructed in the United States will end up costing $70/MWh, but there is significant uncertainty in the construction costs. Investors will likely watch the construction costs of the first few plants built. If they can stay within reasonable budgets and timeframes, the construction of subsequent plants will be encouraged. Two factors could make nuclear more competitive: (1) a federal subsidy of $\sim$ $18/MWh will be granted to the first nuclear plants built; and (2) a proposed carbon tax applied to fossil fuel electrical generation. The subsidy is an extension of the subsidy that already exists for wind power and other renewable sources. The carbon tax would be designed to discourage CO_2 emissions and (hopefully) help mitigate global climate change.

9.5 PROLIFERATION OF NUCLEAR WEAPONS

Since the beginning of the nuclear age, scientists and politicians have tried to separate weapons from nuclear power. Unfortunately, the two are closely related, and this has resulted in nine countries[1] now having nuclear weapons. Highly enriched uranium (>80% ^{235}U) or relatively pure ^{239}Pu are required to build nuclear weapons—fortunately they are difficult to come by. The focus for nonproliferation efforts has been to control access to these materials. The "Megatons to Megawatts" program mentioned above is a good example of a nonproliferation success. Weapons-grade uranium is being converted to nuclear fuel and burned in U.S. reactors.

In addition to taking weapons-grade nuclear material out of circulation, nonproliferation efforts have focused on two key components of the nuclear fuel cycle—uranium enrichment and spent fuel reprocessing—as both can allow production of weapons material. Uranium enrichment facilities can be used to bring the percent ^{235}U up to $\sim$4% for use in a reactor, or the process can easily be continued until the uranium is weapons-grade. Since every nuclear reactor contains ^{238}U, they also produce ^{239}Pu, and any chemical reprocessing of the spent fuel can lead to the isolation of ^{239}Pu. In fact, the U.S. government's once long-standing opposition to the reprocessing of spent fuel was based primarily on nonproliferation grounds. The recent shift to reprocessing unfortunately calls into question its moral leadership in this area.

One key to nonproliferation of nuclear weapons is therefore some sort of control of enrichment and reprocessing facilities. There aren't many of these facilities, but their numbers grew as the nuclear industry rapidly expanded worldwide through the 1970s and early 1980s. If nuclear power is to be expanded again in the first half of the twenty-first century as a way to inhibit global climate change, it is likely new enrichment and reprocessing facilities will be built. Consolidation of these facilities where

---

[1] The United States, Russia, France, Britain, China, India, Pakistan, Israel, and North Korea all have nuclear weapons as of this writing. Iran may also have nuclear weapons in the near future.

redundant, and placement of them under international control has been suggested as a way to ensure they are used only for energy purposes.

Another key is the strengthening or replacement of the United Nation's treaty on the nonproliferation of nuclear weapons (NPT). The NPT went into effect in 1970 and required the nations with nuclear weapons to disarm and prohibited other nations from developing nuclear weapons. In 1970 only five nations (the United States, the United Kingdom, France, the Union of Soviet Socialist Republics, and China) possessed nuclear weapons. Unfortunately, four new nations (India, Pakistan, Israel, and North Korea) have since created nuclear weapons. These four nations are the only nations not currently signatories to the NPT. Equally unfortunate, the five original nuclear powers still have significant stockpiles of nuclear weapons. Some have suggested that the United States begin a program of unilateral disarmament, allowing it to gain some moral leadership while encouraging other nuclear powers to also disarm. They argue that since the end of the Cold War, the nuclear arsenal no longer has anything to deter, so why should it be preserved?

The debate over nuclear power is a complex and sometimes contentious one. The interested reader is encouraged to also look at some of the resources listed in the bibliography. Many contain more detailed discussions of this topic.

9.6 FUSION REACTORS

The proton cycle discussed in Chapter 8 is generally thought to be impractical for fusion power reactors here on Earth, primarily due to its relatively high **ignition** temperature—the temperature needed to start a fusion reaction. Back in Chapter 8, we calculated the temperature needed for ignition of the proton-proton fusion reaction as nearly 13 billion Kelvin! In reality, fusion reactors do not need to get quite this hot, because some reactions will take place at lower temperatures due to the distribution of particle energies at a particular temperature, or because tunneling will occur. At a certain point (several million Kelvin), enough reactions are running that they generate enough heat for the reactions to be self-sustaining. Even so, reaction ignition is a major roadblock in the development of practical fusion power reactors. All of the test reactors built so far require more energy to be put in for ignition than is obtained from the subsequent fusion reactions.

Currently the best hope for a practical fusion reactor lies in the so-called $D + T$ reaction. It is the reaction of deuterium (^2H) and tritium (^3H):

$$^2_1H + {}^3_1H \rightarrow {}^4_2He + {}^1n$$

Each one of these reactions produces 17.6 MeV of energy, and its Coulomb barrier is only 0.68 MeV, thereby minimizing the ignition temperature. The $D + T$ reaction has been successfully performed in a number of research reactors. Unfortunately, in every case, more energy was put into the reactor than was obtained from it. A problem here is sustaining the fusion reaction. For this to happen, fuel needs to be fed into the reactor while products are removed. It is not yet clear how this can be done.

Some fusion enthusiasts like to state that it does not produce radioactive waste. This is not true. Short half-life nuclides with low mass numbers will be formed in any fusion reactor. The fact that they are short-lived means that fusion plant operators will just need to wait a little while for them to decay to stable nuclides. The neutrons produced in the D + T reaction are of greater concern. These neutrons could interact with some of the metals used to build the reactor, making them radioactive through n, γ reactions. For example, cobalt (often used in high-performance metal alloys) would form ^{60}Co, a beta- and high-energy gamma-emitting nuclide with a half-life of 5.3 years.

$$^{59}\text{Co} + {}^1\text{n} \rightarrow {}^{60}\text{Co} + \gamma$$

$$^{60}\text{Co} \rightarrow {}^{60}\text{Ni} + {}_{-1}e + 2\gamma$$

Another big technical challenge for a practical fusion reactor is how to confine the very hot plasma that would be generated. The confinement issue has been solved by very high magnetic fields. At these temperatures the fuel exists as plasma, bare nuclei, and free electrons in a gaseous soup. Since everything is charged, it can be suspended in a magnetic field and will never touch the walls of its container. The optimal shape for this plasma is like a donut (toroid).

Because of the enormous challenges in researching and developing a practical fusion reactor, an international consortium called International Thermonuclear Experimental Reactor (ITER) has been formed to demonstrate the feasibility of fusion as a power source. ITER has finalized a prototype design and hopes to have its first reactor up and running by 2016. The reactor is to be built near Cadarache, in the south of France.

Finally, you probably noticed that fusion produces a lot less energy per reaction than fission. The D + T reaction is one of the best, but only generates 17.6 MeV per reaction, while a typical fission reaction produces ~200 MeV. This may just be a matter of perspective. If we look at how much energy is released in terms of MeV per mass unit of "target" nuclide, fusion rules! Fusion cranks out ~6 MeV/u while fission only produces ~0.8 MeV/u. This argument is only important for weapons, where delivering a payload (mass) costs money and energy—placed on the same rocket, more bang can be generated from a fusion weapon than a fission warhead with the same mass.

9.7 NUCLEAR WEAPONS

9.7.1 Fission Weapons

Unfortunately, an atomic (fission) bomb is a relatively simple device to build; fortunately, getting the fissile material to put into it is rather difficult. The material should have a high probability of undergoing neutron-induced fission (σ_f), produce a fair number of neutrons with each fission reaction (on average), and have a reasonably long half-life—after we go to all the trouble of building a bomb, it'd be nice to be able to store it for a while. Finally the fissile material should be accessible—either something we can dig up or easily make in a nuclear reactor. Three nuclides that fit all the criteria well are ^{233}U, ^{235}U, and ^{239}Pu. ^{235}U occurs naturally, but requires high enrichment (>80%) for use in weapons. ^{233}U and ^{239}Pu need to be made in a reactor, then chemically separated.

Despite intense security surrounding the Manhattan Project, the Soviet Union was able to obtain a great deal of information by recruiting a number of spies from those working on the project. It was only four years after the United States first tested an atomic bomb that the Soviets were able detonate one. No spying is necessary today, as fairly good descriptions of nuclear weapons are posted on the Internet.

The basic concept behind a **fission bomb** is also simple—it is based on the accelerating fission chain reaction that occurs when a supercritical mass of fissile material exists. The idea is to quickly create a supercritical mass, and then try to keep it contained for as long as possible. Making the supercritical mass quickly means that very little of the fissile material undergoes fission (and releases energy!) before the supercritical mass is formed. Keeping it contained means maximizing the number of fission reactions (energy released) before the supercritical mass is blown apart, ending the chain reaction. The result should be a really big explosion.

The simplest fission bomb design is the "**gun-type**" (fig. 9.5). In this weapon, two subcritical masses of ^{235}U are separated in the bomb. One is "shot" (like a gun shooting a bullet) into the other by a chemical (conventional) explosive. When the two masses come together, they form a supercritical mass. To help start the fission chain reaction, a neutron **initiator** is placed where the center of the supercritical mass ends up. The initiator's purpose is to inject a large number of neutrons into the supercritical mass at the very moment it is formed. This can be accomplished with a small amount of polonium–beryllium mixture.

As we saw for ^{238}U at the beginning of Chapter 8, ^{235}U decays via alpha emission and by spontaneous fission. The latter will produce neutrons, but has a very low branch ratio (7.0×10^{-9}%). ^{235}U emits a lot more alpha particles than neutrons—if we could convert some of those alpha particles to neutrons at the moment the supercritical mass is formed, a lot more fission reactions can take place, and a much larger explosion will result. ^{9}Be does just that. It reacts with the alpha particles to form ^{12}C and a neutron:

$$^{9}_{4}\text{Be} + ^{4}_{2}\text{He} \rightarrow ^{12}_{6}\text{C} + ^{1}\text{n}$$

Certain isotopes of polonium (^{208}Po and ^{209}Po) can be mixed with the beryllium to increase the neutron flux, as they also undergo alpha decay. With significantly

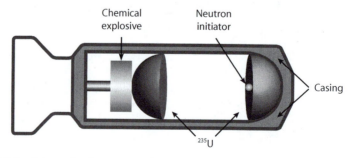

FIGURE 9.5   Schematic of a gun-type fission weapon.

shorter half-lives than ^{235}U, the polonium will have a much higher specific activity, generating more neutrons.

The first atomic bomb used in warfare was a gun-type weapon, called Little Boy, and was dropped by the United States on Hiroshima on August 6, 1945. The Manhattan Project scientists were so confident it would work, they didn't even bother to test a gun-type weapon. Facilitating the decision was the lack of highly enriched uranium—they only had enough to build one bomb. The yield of the bomb was equivalent to 15,000 tons of TNT. Nuclear weapon yield is typically measured in terms of equivalent tonnage of the conventional explosive trinitrotoluene (TNT). For reference, the explosion of one ton of TNT produces 4.2×10^9 J of energy. Little Boy was therefore a 15 kiloton device.

The casing around a gun-type bomb can be made from any sturdy material. Remember, its purpose is to try to keep the supercritical mass together as long as possible (even if it is just a small fraction of a second) to fission as much of the ^{235}U as possible, thereby increasing the weapon's yield. Depleted uranium ($\sim 99.7\%$ ^{238}U) is an excellent choice. Not only is it a strong metal, it will also interact with the fast neutrons produced by the chain reaction in favorable ways. It can reflect some neutrons back into the ^{235}U, and it can undergo fission with others. Both of these types of interactions will increase the yield of the weapon. The proper term for the ^{238}U used in this way is a **tamper**.

The other kind of fission bomb is the implosion-type (fig. 9.6). It is constructed with a hollow sphere of ^{239}Pu, which is surrounded by chemical explosives. The idea is to detonate all of the conventional explosives at exactly the same time, causing the hollow plutonium sphere to collapse to a small solid sphere that is now a supercritical mass. A ^{238}U tamper again surrounds the plutonium sphere to increase the yield.

The Manhattan Project scientists knew that a ^{239}Pu bomb required a different design from ^{235}U. The issue is one of timing. A gun-type weapon does not form the supercritical mass fast enough for ^{239}Pu. Because of its higher σ_f and because it produces more neutrons per fission (on average), ^{239}Pu will begin to blow itself apart before the supercritical mass is completely formed in a gun-type device. Potentially, this could

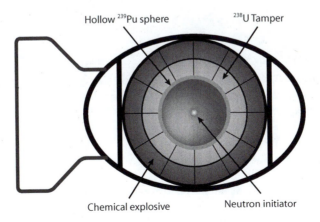

Hollow ^{239}Pu sphere ^{238}U Tamper

Chemical explosive   Neutron initiator

FIGURE 9.6 Schematic of an implosion-type fission weapon.

result in the bomb destroying just itself rather than an entire city. The collapse of a hollow sphere is a faster way to create a supercritical mass. In order for it to work, the timing of the detonations of all the conventional explosives also had to be perfect.

The Manhattan Project scientists were so worried about this, they decided to test an implosion-type device. The code name for the test was "Trinity" and was set up in the Jornada del Muerto (Journey of the Dead Man) Desert in Southern New Mexico. On July 16, 1945, the first nuclear weapon was detonated, ushering in the atomic age. The power of the explosion exceeded all expectations. The awesome power of the blast moved project director J. Robert Oppenheimer to quote the following passage from the sacred Hindu text, the Bhagavad Gita:

> If the radiance of a thousand suns
> Were to burst at once into the sky,
> That would be like the splendor of the Mighty One...
> I am become Death,
> The shatterer of Worlds.

A second implosion-type weapon was dropped on Nagasaki on August 9, 1945, bringing World War II to a horrific conclusion. The Nagasaki weapon was given the name Fat Man, and had a yield of 22 kton.

Fission weapons have been designed with yields ranging from 0.01 to 500 kton. The low end of the spectrum represents an extremely inefficient device—that is, only a fraction of the fissile material undergoes fission and the rest just gets spread over the blast area, or is carried away in the fallout. Little Boy was only 1.3% efficient and Fat Man was ∼17% efficient.

Yield is generally related to the amount of fissile material in the weapon. At a certain point, it is no longer possible to pack more fissile material into a bomb without simply spreading the excess around. In other words, only a certain amount of material can undergo fission before the bomb starts blowing itself apart and the chain reaction shuts down. The upper limit of 500 kton on fission weapon yield is a practical limit. In order to get more destructive power, the very nature of the bomb needs to change.

9.7.2 THERMONUCLEAR DEVICES

Fusion weapons (aka H-bombs or thermonuclear devices) are *much* more powerful. Yields as high as 60,000 kton (that's 60 Mton!) have been made. They are over 4000 times more powerful than the bomb dropped on Hiroshima. As their various names imply, these weapons take advantage the relatively high amount of energy released per unit mass for fusion reactions. As we already know, to get fusion to take place, very high temperatures are needed.

A fission device is used as an **initiator** for fusion weapons (fig. 9.7). It produces the necessary heat, as well as γ-rays and neutrons—all helpful in making fusion possible. The fission device is also referred to as the **primary**, as it is set off first, and its detonation causes the fusion part of the weapon (the **secondary**) to explode.

A **beryllium reflector** is placed behind the primary. Its purpose is to reflect neutrons back into the fissile material of the fission device and toward the secondary.

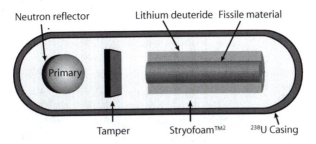

Neutron reflector   Lithium deuteride Fissile material

Primary

Tamper   Stryofoam™²   ²³⁸U Casing

FIGURE 9.7   Schematic representation of a fusion weapon.

Beryllium has a low probability of absorbing neutrons (σ_a), but a relatively high probability of scattering them (σ_s), making it an excellent neutron "mirror." Note that not all neutrons are reflected by the beryllium, it simply has a higher probability of reflecting neutrons than most other materials.

A ^{238}U tamper is placed between the primary and secondary. The fast neutrons produced by the primary can (a) pass right through the tamper, (b) be reflected by the tamper back into the primary, or (c) cause the ^{238}U in the tamper to fission. All of the above are helpful in making a more powerful explosion. The ^{238}U casing also acts as a tamper.

How does the secondary work? As shown in figure 9.7, it is simply a cylinder of fissile material (like ^{239}Pu) wrapped with lithium deuteride (LiD) packed in the device with styrofoam™ ² (and you thought packing peanuts were invented for grandma to ship you fragile stuff!). The tremendous heat and gamma radiation created by the primary turns the styrofoam™ into a super hot plasma, squeezing the secondary. The neutrons that make it through the tamper cause the fissile material inside the secondary to fission, adding more heat ($\sim 10^8$ K!) and pressure to the LiD. Neutrons being released by the fissile material in the primary and secondary react with the lithium to form tritium:

$$^6_3\text{Li} + {}^1\text{n} \rightarrow {}^4_2\text{He} + {}^3_1\text{H}$$

The tritium can now fuse with the deuterium:

$$^2_1\text{H} + {}^3_1\text{H} \rightarrow {}^4_2\text{He} + {}^1\text{n}$$

The neutrons produced are very high energy, and will get a lot of the ^{238}U casing to fission, creating an even bigger explosion. The total yield will depend on how much LiD and fissile material is present. The really high-yield bombs can get rather large and heavy, and would be impractical as a weapon as there would be no easy way to deliver one to a target.

---

² Styrofoam is a registered trademark of the Dow Chemical Company.

Even though fusion weapons use fission triggers, these weapons are often considered relatively "**clean**,"[3] whereas fission nukes, especially those with low yields, are called "**dirty**." Adding certain materials can also make a nuclear weapon dirty. Most commonly, use of ^{238}U as a tamper or casing material will increase levels of radiation in the blast area and the fallout. Likewise, the addition of natural cobalt results in the formation of ^{60}Co. ^{60}Co emits high-energy gamma photons during its decay and has a half-life short enough to give it a high specific activity, but long enough to make it undesirable to move back to contaminated areas for quite a while after the attack. Such a weapon was called a "doomsday device" because it could effectively sterilize a large area for many years. Many forms of popular fiction picked up on this moniker, although the fictional device usually had the ability to destroy the entire planet, something that is not possible with a single real weapon.

9.7.3   Neutron Bomb

The **neutron bomb** (or enhanced radiation device) is a small H-bomb designed for maximum neutron production while minimizing blast and heat, and is therefore considered "clean." This could be done by replacing all of the ^{238}U in the bomb with non-fissile materials. The purpose behind a neutron bomb was to act as a deterrent to an enemy with lots of armored vehicles (the former Soviet Union). The intense neutron radiation would easily penetrate the armor, but be absorbed by the humans inside them. Supposedly this is a great way to disable a large armored invasion near a friendly city. As clean and tidy as they sound, they would still cause a fair amount of local damage, and leave a good deal of radioactive contamination.

9.7.4   Bunker Buster

More recently "**bunker buster**" nukes have been proposed. A nuclear weapon would be placed inside a special casing—designed to hit the ground at high speeds and penetrate the surface to the depth of an underground bunker before detonating. If it worked, such a weapon would be relatively clean in that the surrounding earth would nicely contain the contamination. Some have suggested that such a weapon might be too tempting to use in an otherwise conventional war, which could then escalate the conflict to a full-fledged nuclear war.

9.8   DIRTY BOMBS

Discussion of another (nonnuclear) kind of dirty bomb has sprung up since the terrorist attacks in the United States of September 11, 2001. This dirty bomb is simply some radioactive material added to a conventional (chemical) explosive. Its technical name is "radiological dispersal device" or RDD. The chemical explosive could be as simple as the powder gleaned from fireworks, or as powerful as the fertilizer/

---

[3] It should be a little odd to consider atomic bombs as "clean." They are very deadly and cause unbelievable environmental destruction and human suffering. Additionally they spread their highly radioactive fission products over a broad area. The term is only meant to be relative to other nuclear weapons.

nitromethane mixture used in the 1995 Oklahoma City bombing. Generally speaking, the purpose behind setting off a dirty bomb is not to cause widespread death and destruction, rather it is to cause panic and emotional distress, as well as economic dislocation. It is very unlikely that anyone will die from radiation exposure resulting from a dirty bomb, but it will certainly play into the general public's fear of radiation. It will also contaminate an area that will have to remain evacuated until it can be cleaned up.

QUESTIONS

1. Explain why ^{235}U is better fuel for a nuclear reactor than ^{238}U.
2. ^{233}U has a σ_f of 530 b and a critical mass that is much lower than ^{235}U. What can you conclude about the number of neutrons produced per fission of ^{233}U?
3. Briefly explain what moderators and control rods do in nuclear reactors.
4. Beryllium has a low neutron absorption cross section ($\sigma_a = 0.0076$ b) and a high neutron scattering cross section ($\sigma_s = 5.9$ b). Would it be a good material for a fuel rod, control rod, moderator, and/or coolant in a nuclear reactor? Briefly explain your answer.
5. Beryllium is used as a neutron "mirror" around the core of reactors that need a higher neutron flux. Using the information from Question 4, briefly explain how this works. Is "mirror" a perfect term? Briefly explain.
6. Canadian nuclear reactors are run with naturally occurring uranium, and must therefore use D_2O as a moderator/coolant. Explain why normal water can't be used as a moderator/coolant in Canadian reactors.
7. Explain why enrichment requires a physical rather than chemical process.
8. ^{233}U can be produced in nuclear reactors in a manner similar to the production of ^{239}Pu. What naturally occurring element must be added to a reactor to produce ^{233}U? Write out all nuclear reactions and decays that are part of ^{233}U production.
9. In the event of another nuclear power plant disaster like Chernobyl, state and local politicians have distributed KI tablets to people living near such facilities. The idea is to take the pill and saturate the thyroid with iodide, thereby preventing thyroid uptake of the radioactive iodide released by the plant. Give two reasons why the distribution of these pills could be considered a political placebo, rather than populace protection. Give one reason why distribution of the pills may help protect public health.
10. Assuming all of the radioactive material released by the Chernobyl accident to be ^{137}Cs, what mass (kg) was released?
11. Calculate the percent abundance of ^{235}U on Earth 1.7 billion years ago. Assume the uranium is only composed of ^{235}U and ^{238}U.
12. If you were either for or against nuclear power before reading this chapter, give four to five bullet points outlining the opposite viewpoint. If you had no opinion, flip a coin—heads, write in favor, tails opposed.
13. Do you agree that unilateral nuclear disarmament by the United States would be a positive step for the nonproliferation of nuclear weapons? Explain.

14. The reaction in the problem below has been proposed for use in reactors here on Earth. Its proponents say the only drawback is the lack of ^{3}He. What is the percent abundance of ^{3}He on Earth? Would you expect natural abundance of this nuclide to be low? Explain. The moon is known to have ample supplies of ^{3}He at or near its surface. Why?

15. Based on the energy produced by and the effective Coulomb barrier of this reaction, does it look like a viable reaction for fusion power production?

$$^2_1\text{H} + {^3_2}\text{He} \rightarrow {^4_2}\text{He} + {^1_1}\text{H}$$

16. The most common alpha particles emitted by ^{235}U have an energy of ~4.4 MeV. Is this enough energy for ^{9}Be to be an effective neutron initiator in a fission weapon?

17. What mass of ^{239}Pu was present in the bomb dropped on Nagasaki? What happened to the plutonium that did not fission?

18. Why is a ^{239}Pu fission bomb so much more efficient than one using ^{235}U?

19. Calculate the energy produced by the reaction below, which is part of how fusion weapons detonate. Is a similar reaction also likely for the other naturally occurring isotope of lithium? Briefly explain.

$$^6_3\text{Li} + {^1}\text{n} \rightarrow {^4_2}\text{He} + {^3_1}\text{H}$$

20. The photon torpedoes used in *Star Trek* use antimatter as an explosive. Calculate the mass of antimatter necessary to equal the explosive yield of the bomb dropped on Hiroshima. Comment on whether these weapons could someday become a reality. Why is the word "photon" particularly appropriate to these fictional weapons?

10 Nuclear Medicine

Nuclear medicine is a health-care specialty that uses radioactive nuclides to diagnose and treat disease. Nuclear medicine technologists (NMTs) prepare radiolabeled drugs (**radiopharmaceuticals**), which localize in a specific organ, then obtain images that provide information on the structure and function of the target organ. A wide variety of scans are possible, covering every major organ system in the human body. Nuclear medicine scans are even possible for some animals other than humans. This is a growing field and entire textbooks are devoted to it. This chapter is an introduction to NMT from the perspective of the radionuclide.

10.1 RADIONUCLIDE PRODUCTION

As we've already seen (Chapters 7–9), radioactive nuclides can be prepared using accelerators or nuclear reactors. Traditionally, most nuclides for use in nuclear medicine came from nuclear reactors. There are two ways reactors can prepare these nuclides: (1) as fission products and (2) from neutron bombardment of a material placed in a reactor. To use a fission product, the nuclear fuel must be removed and the desired nuclide must be separated from all of the other components in the fuel pellet. Remember that fission products make up only about 3% of the spent fuel, and hundreds of different fission product nuclides are formed during fission. Additionally the spent fuel is rather toasty (hot and radioactive!), making it difficult to handle. For all of these reasons, obtaining medical radionuclides from spent nuclear fuel is usually undesirable.

A more attractive option for obtaining fission products would be to place a relatively pure ^{235}U target in the reactor. As the neutrons bombard it, its nuclei will fission, providing a much higher percentage of fission products when compared to starting mass (fuel pellets are mostly ^{238}U). ^{99}Mo, ^{131}I, ^{133}Xe, and ^{137}Cs are all commonly obtained this way.

$$^{235}\text{U} + {^1}\text{n} \rightarrow {^{135}}\text{Sn} + {^{99}}\text{Mo} + 2{^1}\text{n}$$

$$^{235}\text{U} + {^1}\text{n} \rightarrow {^{131}}\text{I} + {^{102}}\text{Y} + 3{^1}\text{n}$$

$$^{235}\text{U} + {^1}\text{n} \rightarrow {^{133}}\text{Xe} + {^{100}}\text{Sr} + 3{^1}\text{n}$$

$$^{235}\text{U} + {^1}\text{n} \rightarrow {^{137}}\text{Cs} + {^{97}}\text{Rb} + 2{^1}\text{n}$$

Remember that lots of fission reactions are possible. Also, the reactions above are just examples that happen to lead directly to the desired nuclides. Other reactions that produce more neutron-rich isobars can also decay to produce the desired products.

For example, if ^{99}Nb is formed by a fission reaction, it will quickly ($t_{1/2} = 15.0$ s) decay to ^{99}Mo:

$$^{99}\text{Nb} \rightarrow {}^{99}\text{Mo} + {}_{-1}e$$

Placing targets in a nuclear reactor to undergo an n, γ reaction is also an attractive option. For example, placing ^{50}Cr in a reactor results in the formation of ^{51}Cr:

$$^{50}\text{Cr} + {}^{1}\text{n} \rightarrow {}^{51}\text{Cr} + \gamma$$

Many such reactions are possible, but the main point is that if the target is a pure element or nuclide, then isolating the product should be a lot easier than fishing it out from among all the other stuff present in spent fuel or irradiated ^{235}U. Regardless of how they are prepared, reactor-produced nuclides *tend* to be neutron rich and undergo beta (minus) decay (note that ^{51}Cr is an exception!).

Increasing numbers of medical radionuclides are being prepared with cyclotrons. Cyclotrons are the accelerator of choice because they are relatively compact (about the size of a small car) and affordable when compared to linear accelerators or synchrotrons. Protons, deuterons, and alpha particles are typical projectiles. As mentioned in Chapter 7, ^{18}F can be produced from ^{18}O using a p,n reaction:

$$^{18}\text{O} + {}^{1}\text{H} \rightarrow {}^{18}\text{F} + {}^{1}\text{n}$$

As also mentioned in Chapter 7, it is important to keep the energy requirements of the reaction within the cyclotron's abilities: the target nuclide should be readily available and it should be easy to separate the product. Efficient separation of the product can sometimes be of paramount importance because of the short half-lives of some of the cyclotron-produced nuclides. Some are so short that they are administered directly to the patient from the cyclotron facility. In contrast to reactor-produced nuclides, cyclotron-produced nuclides tend to be proton rich, typically decaying via positron emission or electron capture.

How can a Nuclear Medicine Department in a hospital, clinic, or private practice be assured of a consistent supply of the radionuclides it needs? Both cyclotrons and nuclear reactors are expensive facilities to have around, so most medical radionuclides are purchased and shipped to where they are needed. The farther the production facilities are, the more limited the variety of radionuclides that can be shipped. Those with very short half-lives would not survive the journey if it took too long.

A short half-life is desirable for nuclear medicine, but how do you assure a continuous supply in the absence of a reactor or a cyclotron? In some cases, a shorter half-life nuclide can be produced on-site without the need for a reactor or cyclotron. If a longer-lived parent (to the desired nuclide) nuclide can be produced at the reactor or cyclotron, it may be packaged and shipped to a remote location. As the parent decays, the daughter is gradually produced. If the two can easily be separated, then the daughter can be used for radiopharmaceutical preparations. Such a system is called a **radionuclide generator**.

The parent needs to stay inside the generator while the daughter flows out. This is usually accomplished with an absorbent or ion-exchange material. After the daughter is removed from the generator, time must pass for its activity inside the generator to build back up. Since separation can be repeated many times at regular intervals, the generator is often called a "cow" and the separation procedure "milking." Radionuclide generators are generally examples of transient equilibria and the formulas discussed in Chapter 2 can be used to determine how often to milk the cow.

The most common generator in use today is the ^{99}Mo/^{99m}Tc generator. The ^{99}Mo mentioned above is not produced in a reactor to be used directly in a radiopharmaceutical, but to generate ^{99m}Tc. Figure 10.1 shows the effects on activity by typical milking of the cow. If enough time is allowed to elapse between milking, daughter (^{99m}Tc) activity can build up to close to or above parent activity. Remember from Chapter 2 (equation (2.15)) that maximum ^{99m}Tc (daughter) activity can be calculated from the half-lives of ^{99}Mo (parent, 2.7476 d = 65.94 h) and ^{99m}Tc (6.01 h).

$$t_{max} = \left[\frac{1.44 t_{1/2(^{99}\text{Mo})} t_{1/2(^{99m}\text{Tc})}}{\left(t_{1/2(^{99}\text{Mo})} - t_{1/2(^{99m}\text{Tc})} \right)} \right] \times \ln \frac{t_{1/2(^{99}\text{Mo})}}{t_{1/2(^{99m}\text{Tc})}}$$

$$= \left[\frac{1.44 \times 65.94 \text{ h} \times 6.01 \text{ h}}{65.94 \text{ h} - 6.01 \text{ h}} \right] \times \ln \frac{65.94 \text{ h}}{6.01 \text{ h}}$$

$$= 22.8 \text{ h}$$

This time interval is quite convenient for us humans, as it is very close to one day. In figure 10.1, ^{99m}Tc is removed every 24 h for the first three days, shortly after the ^{99m}Tc activity has peaked. Later in the third day, additional doses are necessary, so the generator is milked again. If insufficient time between milkings elapses, daughter

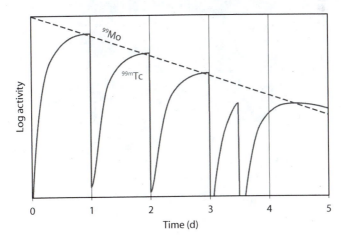

FIGURE 10.1 ^{99}Mo (parent, dashed) and ^{99m}Tc (daughter, solid) activities when ^{99m}Tc is removed at occasional intervals.

activity will still be below parent activity. The generator is not milked on the fourth day, and transient equilibrium will finally be reached late on the fifth day if the generator is not milked again (>7 daughter half-lives).

Note also that parent (^{99}Mo) activity also decreases with time in figure 10.1, as expected for transient equilibrium. Eventually (typically a week), the generator will no longer be able to produce sufficient ^{99m}Tc activity to provide the doses necessary and a new generator will be purchased.

A cut-away view of the ^{99}Mo/^{99m}Tc generator is shown in figure 10.2. Overall, the generator is a cylinder, about a foot tall. Sterile saline solution is added through a port on the top. This solution flows through an Al_2O_3 (alumina) column with the ^{99}Mo packed in at the top. Before loading it on the alumina, it is oxidized to MoO_4^{2-} (molybdate, probably using H_2O_2). TcO_4^- (pertechnetate) is produced through beta decay, and does not bind to the alumina as strongly as molybdate. The relative affinity for alumina can be related to ionic charge. Molybdate has a $2-$ charge, while pertechnetate only has a $1-$ charge. Generally speaking, the higher the charge, the more likely it'll stick to alumina. This allows elution of pertechnetate using the saline solution, without bringing much, if any, parent along. It is usually eluted directly into a **kit**—a bottle containing chemicals that will react with pertechnetate to form a radiopharmaceutical. Other radionuclide generators used in nuclear medicine include: ^{68}Ge/^{68}Ga, ^{82}Sr/^{82}Rb, and ^{113}Sn/^{113m}In.

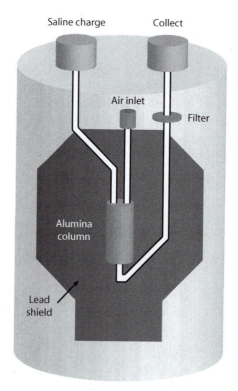

FIGURE 10.2   Cut-away view of a ^{99}Mo/^{99m}Tc generator.

Nuclides that are produced without the presence of isotopes or isomers (carrier) of the desired nuclide are called **carrier-free**. In reality, most radionuclide preparations will result in the presence of some carrier. Therefore, most radionuclide preparations are termed as **no carrier added** (NCA) to indicate that no stable isotopes or isomers of the product were deliberately added. The ^{99m}Tc preparation described above may initially seem carrier-free, but as soon as it is produced, ^{99m}Tc begins to decay to ^{99}Tc. Some ^{99}Tc (carrier) will be present in any sample of ^{99m}Tc. The presence of carrier is undesirable because isotopes or isomers of the desired nuclide will have different nuclear properties, and will therefore be unlikely to help in imaging the target organ. Carrier generally refers to a stable isotope or isomer, but a radioactive isomer like ^{99}Tc is also considered a carrier because it is (very nearly) a purely beta-emitting nuclide with a very long half-life (213,000 years!).

10.2   RADIOPHARMACEUTICALS

As mentioned at the beginning of this chapter, a radiopharmaceutical is a drug with a radioactive nuclide used for diagnostic or therapeutic purposes. In reality, $\sim$95% of all radiopharmaceuticals are used for medical diagnoses, and only $\sim$5% are used for therapy. Radiopharmaceuticals can be as simple as a nuclide (^{133}Xe), ion (^{125}I$^-$ or ^{82}Rb$^+$), or a small organic (^{18}F-glucose) or inorganic (^{99m}Tc(CNR)$_6$$^+$) compound; or they can be as complex as a large biomolecule (radiolabeled protein, antibody, etc.). The exact nature of the radiopharmaceutical depends on how it will be used.

What are the qualities of a good diagnostic radiopharmaceutical (sometimes called a radiodiagnostic agent)?

It should have a good biodistribution. In other words, it should localize primarily in the target organ. It is also helpful if it does not localize in other organs nearby, as this may interfere with obtaining a good image of the target.

The radionuclide should emit only gamma rays. Gamma emissions are important, as a nuclide emitting solely alpha or beta particles will not be visible outside of the patient, and will give the patient a significant internal dose. If it decays via IT, it should have a low conversion coefficient (α_T). If it decays via EC, it should have a high fluorescent yield (ω_T). Additionally, the gamma rays should be between 50 and 500 keV in energy. Any lower, and the patient will absorb too many, and any higher will make detection too inefficient (see Chapter 6). The amount of radionuclide administered should be enough to provide a good image, but no more—the dose to the patient needs to be minimized.

The radioactive nuclide should also have a short half-life. The half-life needs to be long enough for the radiopharmaceutical to be prepared, injected into the patient, localize in the target organs, and be imaged. As a practical matter, it should be at least a few minutes. A short half-life will maximize the amount of decay that takes place during imaging, ensuring good pictures. It will also minimize the dose to the patient (and others nearby!) and possible contamination (due to excretion) following the procedure.

Finally, the nuclide should be readily available, relatively pure, and decay to a stable, benign daughter. As discussed earlier, the radionuclide needs to be produced on-site or shipped to the site. A radioactive daughter will mean additional patient dose and may interfere with the scan. Benign simply means that it should not be harmful to the patient and the environment—the radionuclide should not decay to an element that is highly toxic for humans or particularly harmful to the environment. These are not great concerns, as very small amounts of radioactive materials are used in diagnostic procedures.

The radiopharmaceutical should be relatively inexpensive and easy to make. If the materials (radionuclide + chemicals) to make a radiopharmaceutical are too costly, it is less likely to be approved by insurance companies (the kiss of death!). If its synthesis is too complex, radiopharmacies are less likely to be able to make it.

Once imaging is complete, the radiopharmaceutical should be metabolized and excreted by the patient in a reasonable amount of time. This also minimizes the patient dose. If the patient will excrete a significant amount of the dose soon after the procedure, then a special bathroom should be used. The patient should also be warned if their urine and/or feces represent any radiological hazard.

^{99m}Tc is the most commonly used radionuclide for radiopharmaceutical preparation. Its generator is convenient in that ^{99}Mo is easily prepared in a reactor and the half-lives of ^{99}Mo and ^{99m}Tc lend themselves well to human schedules (as shown above). Also noted above is that the Mo–Tc generator has a useful lifetime of about one week, also convenient for institutions working a five-day week. The only significant disadvantage to ^{99m}Tc is that it does not decay to a stable daughter.

^{99m}Tc generators are also relatively inexpensive, and ^{99m}Tc emits a 140 keV gamma photon in 89% of its decays (low α_T) with a half-life of 6.01 h. Technetium is an element with a great deal of chemical flexibility. It can form a wide variety of chemical compounds—with coordination numbers (number of things attached to the metal) around Tc from 4 to 9 and oxidation numbers (relative number of electrons localized on the metal) from -1 to $+7$. ^{99m}Tc is typically produced in the generator as TcO_4^- (pertechnetate ion, coordination number $= 4$, oxidation number $= +7$) and reduction (lowering of the oxidation number) is almost always required. This is accomplished with the tin(II) (Sn^{2+}) ion, which is oxidized to tin(IV) (Sn^{4+}). When the technetium is reduced, other chemicals (**ligands**) must bind to it. For example, the preparation of the myocardial perfusion agent ^{99m}Tc-Sestamibi can be characterized by the following chemical reaction:

$$TcO_4^- + 6CNR + 3Sn^{2+} + 8H^+ \rightarrow Tc(CNR)_6^+ + 4H_2O + 3Sn^{4+}$$

where CNR is 2-methoxy isobutyl isonitrile, the ligand. $Tc(CNR)_6^+$ has a coordination number of 6 and an oxidation number of $+1$.

The chemistry is easy to write down on paper, but there are some challenges when preparing radiopharmaceuticals. First, the solutions are very dilute—$^{99m}TcO_4^-$ has a concentration of about 10^{-9} mol/l. The reactions must also proceed in high chemical yield, as injection of other radioactive chemical species will likely interfere with imaging of the target organ.

$Tc(CNR)_6^+$ has a rather symmetric structure, and appears to the body as a large positively charged sphere. The body confuses it for potassium ion, so a fair bit of it ends up in the heart, making it a good myocardial imaging agent. Other radiopharmaceuticals used for heart imaging look even more like the potassium ion (K^+); they are $^{82}Rb^+$ and $^{201}Tl^+$.

Sometimes the physical form of the radiopharmaceutical is more important than its chemical form. A couple of ^{99m}Tc colloids are used for different imaging applications. A colloid is a bunch of molecules that stick together or a really large molecule. Colloids are still small enough to disperse, but not dissolve, in a solvent, but not big enough to precipitate out. Generally speaking, a colloid is the dispersal of one phase of material in another. Smoke is a sold dispersed in gaseous air, and is therefore a colloid. Milk is also a suspension of liquid fats in liquid water—the fat is a colloid. The subsequent mixture is called an emulsion because its two components are immiscible. The key to a ^{99m}Tc colloid is the size of the multimolecular aggregate that forms, as the size of the colloid will determine its application.

Most ^{99m}Tc radiopharmaceuticals in current use can be related to ^{99m}Tc-Sestamibi in that they started out as a compound that was found to be easily prepared from nanomolar solutions of pertechnetate, were stable under physiological conditions, and have some (typically unanticipated) favorable biodistribution. Most were discovered through trial and error, which can be a time consuming and inefficient way to discover new radiopharmaceuticals. More recently, radiopharmaceutical research has started with biodistribution, then looked for a way to attach a radioactive nuclide. Two successful approaches are discussed below.

The first involves forming a chemical linkage between a molecule with a known biodistribution, such as an antibody, carbohydrate, enzyme, protein, or peptide. In the example shown in figure 10.3, a diethylenetriamine pentaacetate (DTPA) is wrapped around an indium(III) ion (^{111}In^{3+}). When a single ligand binds to a single metal ion with more than one group, it is called a **chelate**. Chelating ligands like DTPA can also be used to help sequester radioactive materials that have been inadvertently ingested. The DTPA wraps itself around the radioactive ion and facilitates its clearance from the body. Back to the reactions below, notice that one of the five acetate groups remains unbound, and is used to form a chemical bond to the biomolecule. This approach could be called "a tetherball," where the tether is the biomolecule and the ball is the chelated metal ion.

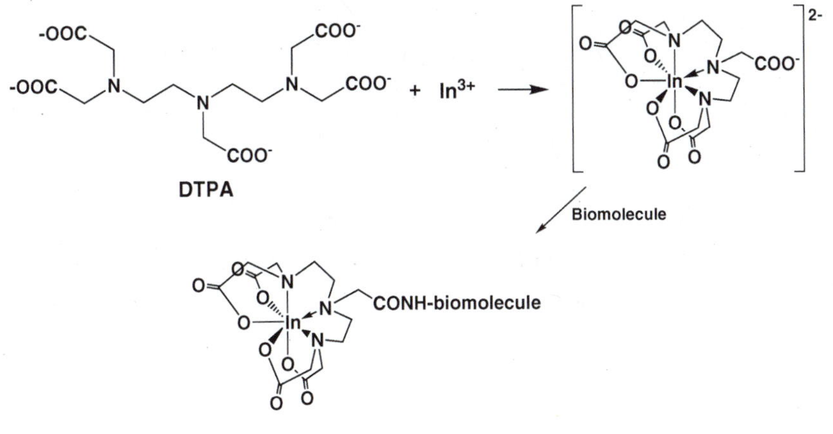

FIGURE 10.3   Preparation of a "tetherball" radiopharmaceutical using indium and DTPA.

This approach can fail if the presence of the ball affects the tether's ability to localize in the desirable organ. It could be that the ball binds to a critical part of the tether or that the addition of the ball changes the overall shape of the biomolecule to a significant extent. Synthesis can also be more complex if the two steps need to be separated.

The most elegant approach is to incorporate the radioactive nuclide in the biomolecule—replacing one of the atoms with a radioactive isotope. For example, replacing the nitrogen in an amino acid (a basic biochemical building block!) with a ^{13}N. The synthetic challenges here are daunting, especially since most of the radio-active nuclides that could be used have rather short half-lives and are cyclotron-produced. ^{18}F-fluorodeoxyglucose (FDG) is probably the most widely used of this type of radiopharmaceutical, although it represents a minor biochemical compromise.

As can be seen in figure 10.4, the compromise is that FDG replaces one of the hydroxyl (−OH) groups with a fluorine (−F). In terms of size and arrangement of electrons, these two groups are similar enough that the body mistakes FDG for glu-cose. FDG is used to study metabolism in the brain and heart, and is used to detect tumors at the cellular level.

Flourine-18 is a positron-emitting nuclide; therefore FDG is used as a positron emission tomography (PET) agent. As you remember from Chapter 4, positrons annihilate with electrons to produce two 511 keV photons that travel on trajectories that are exactly180° apart. If a patient is injected with some FDG, it will be emitting positrons, which travel a short distance before annihilating, generating two photons, 180° apart. PET works by passing the patient through a circle of detectors.

The chief advantage of PET over using a nuclide like ^{99m}Tc is that PET generates two photons per decay, whereas ^{99m}Tc only produces a single photon to be detected. Both can provide 3-D images of the target organ provided detectors can (more or less) encircle the patient. When this is done with a single photon nuclide like ^{99m}Tc it is called single photon emission computed tomography (**SPECT**).

PET imaging is growing in importance within nuclear medicine, in part because radiopharmaceuticals such as FDG are examples of "**molecular imaging**." That is to say, they have the ability to illuminate biological processes at the molecular or cellular level. They allow understanding of physiology through the actual biochemi-cal reactions that take place inside human beings. Since these processes are some-what unique to each individual, this has the effect of personalizing the diagnosis and treatment of disease. Molecular imaging is so prominent for nuclear medicine that the U.S. professional organization for nuclear medicine, the Society of Nuclear

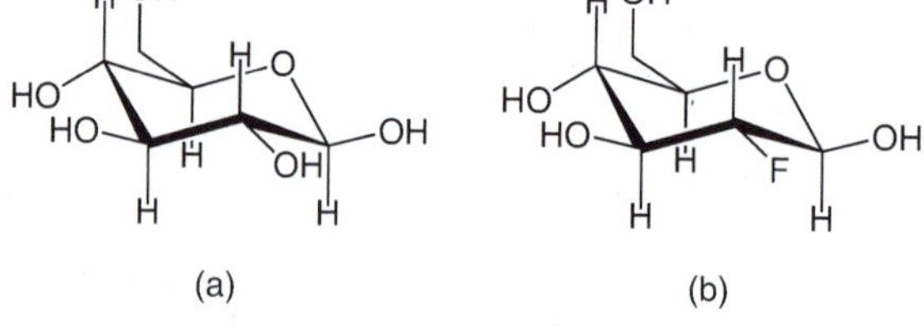

FIGURE 10.4   (a) Glucose and (b) FDG.

Medicine (SNM), expanded its mission in 2007 to include "advancing molecular imaging and therapy." As our understanding of human beings continues to become more detailed at the biochemical level, our ability to detect and treat disease at this level will become more effective.

Coupled with the increasing abilities of radiopharmaceuticals to detect disease at the cellular and molecular level are **fusion** technologies. Not to be confused with nuclear fusion, this type of fusion involves the coupling of nuclear medical imaging with computed tomography (CT) or magnetic resonance imaging (MRI). The nuclear medicine scan provides physiological information while CT or MRI provides anatomical information. When the two scans are put together, they provide medical professionals with detailed information within and around the volume of interest. To accomplish both scans, the immobilized patient is typically placed on an exam table that travels through both types of scanners. For example, a PET/CT camera will pass the patient through a PET detector, then a CT scanner, like poking your finger through the holes of two doughnuts sitting up on their sides (fig. 10.5).

So far, we've only discussed diagnostic agents. As mentioned above, nuclear medicine is moving toward molecular imaging *and therapy*. As molecular imaging advances, it is easy to imagine swapping a gamma- or positron-emitting nuclide with one that emits alpha or beta particles, Auger electrons, or low-energy γ-rays. These forms of ionizing radiation will deposit a lot of energy (i.e., do a lot of damage!) in nearby tissue. Only specific tissues (like tumors) are targeted for destruction, so the radiotherapy agent will need an excellent biodistribution. Like radioimaging agents, they should also be readily available and easy to make.

At the time of this writing, molecular radiotherapy remains an active area of research, but no such agents are currently in common use. Part of the problem is that the agents do not localize strongly or quickly enough in the target tissue, leading to a significant dose to healthy tissue. It should also be noted that beta- and low-energy

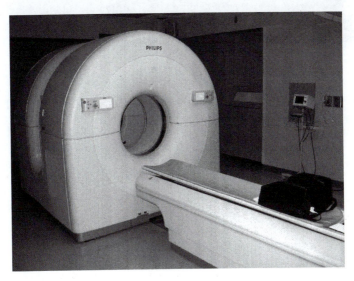

FIGURE 10.5 A Phillips PET-CT camera.

gamma-emitting nuclides are currently showing the most promise. The daughter recoil that follows alpha emission tends to eject the daughter nuclide from the radiopharmaceutical, and likewise, Auger electron emission tends to be disruptive to the chemical bonds that hold the daughter to the rest of the radiopharmaceutical.

The only radiotherapy agent in common use today is a simple ion: $^{131}I^-$. As mentioned earlier in this chapter, ^{131}I can be separated from the other fission products of ^{235}U, but it can also be produced from the n,γ reaction production of ^{131}Te and its subsequent decay.

$$^{130}Te + {}^1n \rightarrow {}^{131}Te + \gamma$$

$$^{131}Te \rightarrow {}^{131}I + {}_{-1}e$$

Iodide collects in the thyroid, so $^{131}I^-$ is used to treat thyroid cancer and hyperthyroidism. ^{131}I also emits γ-rays as part of its beta decay, so it can also be used for imaging. This is also handy when used for therapy because the biodistribution can be determined as well as the dose received by the thyroid. ^{131}I decays to the stable and chemically inert nuclide, ^{131}Xe.

10.3 RADIOTRACERS

Radiopharmaceuticals are sometimes referred to as radiotracers, or simply as tracers. These are more general terms that can apply to experiments using small amounts of radioactive nuclides to "follow" any chemical or biological process. The basic idea is that a radioactive nuclide will exhibit the same chemical and physical behavior as its stable isotopes. By mixing in a small percentage of radioactive nuclide into a system under study, a lot can be learned about it without changing it in any way. Examples include a wide variety of analysis, including determining the solubility product constant (how much dissolves, K_{sp}) of relatively insoluble salts or understanding how CO_2 is assimilated in plants during photosynthesis.

QUESTIONS

1. The two fission reactions given early in this chapter are reproduced below. Would you expect these reactions to have high fission yields? Briefly explain.

$$^{235}U + {}^1n \rightarrow {}^{135}Sn + {}^{99}Mo + 2{}^1n$$

$$^{235}U + {}^1n \rightarrow {}^{131}I + {}^{102}Y + 3{}^1n$$

2. A 28 mCi dose of a ^{99m}Tc myocardial agent is injected into a patient. How many atoms of ^{99m}Tc were injected? What mass of ^{99m}Tc is present? This radiopharmaceutical kit also contains 0.050 mg of $SnCl_2 \cdot 2H_2O$. What is the mole ratio of tin to technetium?

3. A ^{99}Mo/^{99m}Tc generator currently has no ^{99m}Tc activity. How long will it take for ^{99m}Tc activity to peak?

4. What would be a simple, safe, and expeditious way to test for ^{99}Mo impurities in a ^{99m}Tc sample recently eluted into a radiopharmaceutical kit from a ^{99}Mo/^{99m}Tc generator?

5. Would ^{60m}Co be a good choice for a nuclear medicine diagnostic procedure? Explain.

6. What are the differences between a diagnostic and a therapeutic radiopharmaceutical?

7. Why are ^{82}Rb$^+$ and ^{201}Tl$^+$ good myocardial imaging agents?

8. Why would radioactive nuclides produced by cyclotrons be preferred over the same ones produced in a nuclear reactor? When would those from reactors be preferred?

9. Why are PET images inherently fuzzy?

10. What is the purpose of a radionuclide generator? How does it relate to dairy farming?

11 Radiation Therapy and Food Irradiation

Radiation therapy is the medical application of ionizing radiation to treat disease. Food irradiation is the use of ionizing radiation to kill biological organisms that are deleterious to foods. It may seem strange to group these two topics together in a single chapter, but they have a lot in common. Both typically use high-energy photons to kill the bad stuff (diseased cells or bacteria) while trying to leave the bulk (healthy tissue or edible food) unaffected.

11.1 RADIATION THERAPY

Today, most radiation therapy (RT) procedures involve the bombardment of tumors with high-energy X-rays. The remaining RT procedures involve the placement of a sealed radioactive source inside or near diseased tissue with the intent to kill it. The former are generally termed external beam therapy and the latter implant therapy (or brachytherapy). As mentioned above, the goal is to deliver the greatest dose to a tumor while minimizing dose to the surrounding (healthy) tissue. The ratio of abnormal cells killed to normal cells killed is called the **therapeutic ratio**. These ratios are generally high for tumor treatment because tumor cells are rapidly dividing, and therefore are much more susceptible to radiation damage. The history of RT is all about increasing the therapeutic ratio.

The first 50 years (1895–1945, or so) of external beam therapy were dominated by treatment using machines that could only generate photons up to 500 keV. Because these instruments generated X-rays in much the same fashion as diagnostic X-ray machines (used to look for a broken bone …) they are frequently called "conventional" X-ray machines. These treatments were often limited by skin reactions because the maximum dosage was always delivered to the skin. In fact, treatment time was often determined by observation of the skin under the X-ray beam—when the skin started getting too red, treatment was terminated.

Implant therapy was fairly popular in those days. Back then, the major radioactive sources were members of the naturally occurring decay series. ^{226}Ra was commonly used as it was easily separated from uranium ores. As you will recall (Chapter 1), ^{226}Ra is in the middle of the ^{238}U decay series—there are still a dozen or so radioactive nuclides in the series before it stops at ^{206}Pb. This is bad news, as it means that all these other nuclides will be present in any sample of ^{226}Ra once it has sat around for a while (about a month). All kinds of different ionizing radiation will be emitted. Enough high-energy gamma radiation was emitted that those working with these sources on a daily basis couldn't help but be exposed to significant doses.

Alpha decay is a big part of the decay of ^{226}Ra to ^{206}Pb. Every one of these alpha particles becomes a helium atom inside the sealed source. Pressure will build up from the helium and the ^{222}Rn produced in the decay chain and eventually burst the container. This was a big problem with radium implant sources because the radium was in the form of a fine powder—when a sealed source burst it created quite a mess to clean up.

Finally, it was tough to get these sealed sources to stay put. Our bodies move—even at rest we breathe, circulate blood, and digest. It is very difficult, even for the most disciplined person, to remain motionless for long. When we move, something placed in our tissues will also move—not always in the same way. As a result it was often difficult to ensure adequate dose distributions from sealed sources placed inside humans. For all the reasons mentioned above, implant therapy fell out of favor during the first half of the twentieth century.

This was also a time when the biological effects of ionizing radiation were often misunderstood. Unfortunately there were some who took advantage of this lack of understanding and sold bogus radiation "treatments." Radium salts (inorganic compounds) were sold in health elixirs. People paid for mine tours to breathe air containing relatively high concentrations of radon or to soak in waters containing radioactive ions. Radium-containing cosmetics and radium painted clock and watch dials were also popular in the early part of the century because they glowed all night long. For a good part of the twentieth century, uranium compounds were used to provide colors in the glazes of ceramic products, most famously in orange Fiesta®[1] dinnerware for a few years.

After World War II, nuclides like ^{60}Co (1.17 and 1.33 MeV γ-rays) became available, and radiation therapy instruments using ^{60}Co sources became the mainstay for ~30 years. As a result, some aspects of being a radiation therapist became more complicated. Higher-energy photons meant that the maximum dose was now received below the skin, improving the therapeutic ratio but eliminating skin damage as a way to determine treatment times.

It was also during this time that radiation therapists started repositioning the γ-ray beam during treatment or between treatments—firing at a tumor from different angles, as shown in figure 11.1. This is known as **isocentric** therapy because the tumor is located at the intersection (isocenter) of the various beam paths. Isocentric therapy minimizes the dose to the healthy tissue between the tumor and the source by changing the path the photons take to zap the tumor. This technique greatly increases the therapeutic ratio. Determining variables such as treatment time, optimal beam shape, and angles of entry became quite complex when using isocentric techniques. Fortunately, computers were also developed in this time period and greatly facilitated treatment planning.

In the late 1960s, researchers at Los Alamos National Laboratory developed compact linear accelerators, which were capable of generating >1 MeV X-rays in a relatively small space (like a hospital room). Once again, higher-energy beams meant the ability to deliver the maximum dose deeper inside the patient, improving the therapeutic ratio. As a result, these megavoltage machines (commonly called "linacs") started replacing ^{60}Co instruments in the 1970s. It took a while, but they are now pretty much the only external beam therapy machines used by radiation

---

[1] Fiesta is a registered trademark of the Homer Laughlin China Co., Newell, West Virginia.

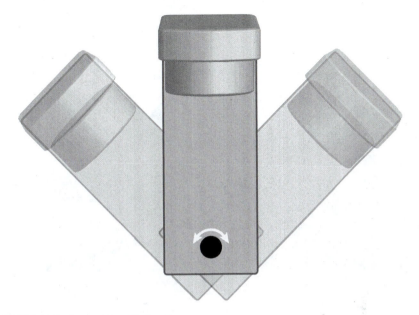

FIGURE 11.1   Isocentric radiation therapy.

therapists (fig. 11.2). With the dawning of the megavoltage era, still greater complexity was introduced. These are pretty high-tech instruments that produce very intense, very high-energy, very sharp beams. They can also be moved with greater flexibility relative to the tumor, and some can even change the shape of the X-ray beam as the beam is moved around the patient. Treatment planning is now heavily dependent on computers, but therapeutic ratios are higher than ever.

Implant therapy has experienced a bit of a renaissance in recent years, thanks in part to the use of new nuclides such as ^{192}Ir and ^{103}Pd. Iridium and palladium are chemically inert, and therefore require minimal encapsulation. Other implant sources were typically encapsulated in stainless steel, while ^{192}Ir and ^{103}Pd can be encased in thin plastic. This allows more ionizing radiation (low-energy gamma photons and beta particles) to penetrate the casing. ^{192}Ir and ^{103}Pd both emit lower-energy photons, and ^{192}Ir also emits beta particles. Because of their high specific activities and minimal encapsulation requirements, these sources can be quite small. In fact, they are often referred to as "seeds." Implant sources can be placed permanently inside a patient, but they are usually only temporarily put inside someone. To accomplish this they can be placed on a wire or plastic line and inserted into the patient through a catheter.

Electron and proton beams are sometimes used for RT. Electron beams are readily available as the same X-ray machines used by radiation therapists generate a beam of high-energy electrons as part of their X-ray production. Instead of hitting a target to generate X-rays, these electrons can be allowed to hit the patient instead. Proton beams are a little harder to come by, as they need a cyclotron to accelerate them, and rather large and powerful magnets are required to bend the beam.

Since both electrons and protons are charged particles, they tend to deposit a large proportion of their energy at a specific depth below the surface of the matter

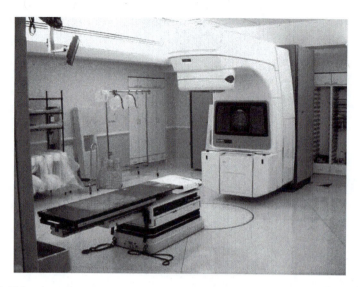

FIGURE 11.2   A modern-day linear accelerator radiation therapy machine. (Photo courtesy of Kris Saeger.)

(the Bragg peak—see Chapter 5). This depth depends on the energy of the particle; the greater the energy, the greater the penetration. This means that electron and proton beams tend to have great therapeutic ratios. The downside is that, as charged particles, they can undergo scattering by the nuclei of the matter they penetrate (more from Chapter 5), causing the beam to spread a bit inside the patient. They also have a much greater probability of interacting with matter than high-energy photons and therefore cannot always penetrate as deeply as needed. Despite their drawbacks, both have specific applications where they outperform high-energy photon beams or implants, so both experience some use.

Beams of other particles have been examined for RT, but none have found a practical application. Beams of neutrons, pions (aka π-muons), alpha particles, and nuclei of ^{20}Ne and ^{40}Ar have all been examined.

As mentioned in the previous chapter, nuclear medicine is currently researching molecular radiotherapy—designer molecules with a very specific biodistribution and a radioactive payload. If successful, these therapies might be well suited to conditions that cannot be effectively treated with current beam or implant techniques. One example would be a cancer that has metastasized to other areas of the body. Beam and implant therapies work best on tumors that are well localized—if one is spread throughout a vital organ or the body, it is difficult to treat with these techniques without destroying the organ or body. Ionizing radiation delivered on the cellular or molecular level could be much more effective.

11.2   FOOD IRRADIATION

The U.S. Postal Service uses ionizing radiation to kill anthrax spores or other potentially harmful organisms that could be sent through the mail. High-energy photons are typically used because of their ability to penetrate packaging. Ionizing radiation is also widely used to sterilize medical supplies like bandages and consumer products such as tampons and condoms. A common public misconception is that irradiation will make their mail, bandage, or condom become radioactive. We learned in Chapters 5 and 7 that this can't be true—only projectiles such as neutrons, protons and other nuclei could cause residual radioactivity. Most likely, ionizations will be caused and chemical bonds will be broken—hopefully killing the bacteria and/or bugs while leaving the rest of the irradiated object alone. Just like RT, we want to kill the bad stuff, while minimizing damage to the rest.

Food irradiation uses X- or γ-rays, or electron beams, to rid food of pests or bacteria so that the food will not be damaged or spoil as quickly. In some cases it can serve as an effective substitute for chemical pesticides and preservatives. It only works well with certain foods, and is typically used on fruits, vegetables, and spices. In some cases it can dramatically increase the shelf life of these products. Food irradiation can be used with meats and could be an effective method to prevent *E. coli* contamination in ground beef. Unfortunately, irradiation of meats can leave it with a bad taste—some say it is like burnt hair, disgusting!

Because very high doses are used, we can't expect that the ionizing radiation will only target the bugs. Some chemical damage is also done to the food. In the case of meat, it clearly alters some of the chemicals that give it flavor. Opponents to food irradiation suggest that irradiation affects the chemical makeup of any food, and should not be used. Attempts to detect these chemicals have largely proved fruitless, which is expected since these chemicals could only be formed in very low quantities. The important question to ask is which has the most benefit and the least risk: treatment with chemicals, treatment with ionizing radiation, or no treatment at all. There are risks and benefits to each.

A typical layout for an irradiation facility is shown in figure 11.3. The stuff to be zapped is loaded on a conveyer belt that takes the product into the irradiation room then out again to be distributed. The irradiation room has extra-thick walls and contains a high-activity source (typically ^{60}Co or ^{137}Cs) or a radiation-generating device (for an X-ray or electron beam). The energy and type of the ionizing radiation as well as the size and materials of the packaging must all be considered to ensure that all

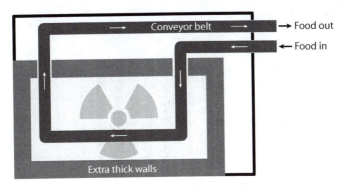

FIGURE 11.3   A schematic for a food irradiation facility.

of the package contents receive the required dose. When not in use, the radioactive source can be lowered into a shielded storage area or switched off.

QUESTIONS

1. How are RT and food irradiation alike?
2. Calculate the specific activity of ^{192}Ir. Why is high specific activity good for an implant source?
3. Why do electrons tend to spread out their dose when used in therapy, when photons do not?
4. What is the most likely interaction between high-energy photons ($\sim$10 MeV) and the low-Z matter that makes up our bodies?
5. The dose delivered to RT patients undergoing external photon beam therapy is due primarily to the photons interacting and what else?
6. Describe a simple way that dose at some depth inside a human could be estimated for external beam photon therapy.
7. A patient has a small tumor in his neck. It is only 2 cm below his skin, but his spine is not far behind the tumor. The spine is very sensitive to ionizing radiation, so it is important to minimize its dose. Would it be better to treat this tumor with photons or electrons? Briefly explain.
8. Calculate the activity for an ^{192}Ir seed that was calibrated 200 days ago at 9.45 Ci.

12 Radiation Protection

Ionizing radiation is frightening, in part because the risks of exposure are expressed as probabilities. If a person dies of cancer it is impossible to point to an exposure to ionizing radiation 20 years in that person's past and state with absolute certainty that the cancer was a result of that exposure. Despite this uncertainty, radiation hazards and risks can be understood and controlled. This chapter will help you understand how ionizing radiation interacts with humans, when to get excited about it, and what to do about it. This chapter builds on what you've learned about the detection of ionizing radiation (Chapter 6) and its interactions with matter (Chapter 5).

12.1 TERMS

It is important to distinguish between exposure and dose. **Exposure** is the amount of ionizing radiation striking an object. **Dose** is the amount of energy absorbed by an object as a result of radiation exposure. As humans, we are exposed to a lot of radiation in the form of γ-rays, which are part of cosmic radiation. Because these γ-rays are high-energy photons, their odds of interacting with us (we are made up of mostly low-Z atoms!) are relatively low. Therefore our *dose* from cosmic radiation is low, even though our *exposure* is high.

As noted back in Chapter 1, the biggest dose to humans living in the United States comes from radon (^{222}Rn). Radon is in the air we breathe and in the water we drink (and it has always been there!), so we ingest a fair bit. ^{222}Rn is an alpha-emitting nuclide, with a relatively short (3.8 d) half-life—so there's probably some inside of us right now, decaying away. Those alpha particles will deposit all of their energy inside our bodies, since alpha particles have high specific ionization (SI) and linear energy transfer (LET) values and very short ranges. Even though it is the source of our largest dose, it is not because we are exposed to relatively large quantities of radiation from ^{222}Rn, it is because the energy of the decay is rather efficiently transferred to our bodies. Remember that exposure and dose are different—exposure is all ionizing radiation around something, but dose will depend on the type of radiation and the nature of the matter it interacts with.

Exposure can be more strictly defined as the amount of charge (ion pairs!) delivered to a mass (or volume) of air. It has SI units of coulombs per kilogram (C/kg). It sometimes is expressed in units of Röntgen (R). One Röntgen produces about 2×10^9 ion pairs in 1 cm^3 of air:

$$1\,R = \frac{2.58 \times 10^{-4}\,C}{kg}$$

Likewise, dose can be defined as the amount of energy delivered per unit mass. The SI unit for dose is the gray (Gy), which is equal to a joule of energy dumped into 1 kg of matter:

$$1 \text{ J/kg} = 1 \text{ Gy}$$

In the United States, units of rad (*r*adiation *a*bsorbed *d*ose) are commonly used. It is defined as 100 erg/g:

$$1 \text{ rad} = \frac{100 \text{ erg}}{\text{g}} = \frac{10^{-2} \text{ J}}{\text{kg}} = 1 \text{ cGy}$$

Example: How much energy (MeV) is deposited in 91 kg by a 1.00 Gy dose?

$$1.00 \text{ Gy} = \frac{1.00 \text{ J}}{\text{kg}}$$

$$\frac{1.00 \text{ J}}{\text{kg}} \times \frac{\text{MeV}}{1.602 \times 10^{-13} \text{ J}} \times 91 \text{ kg} = 5.7 \times 10^{14} \text{ MeV}$$

What is this dose in rad?

$$1.00 \text{ Gy} \times \frac{\text{rad}}{10^{-2} \text{ Gy}} = 100 \text{ rad}$$

We know that ionizing radiation interacts differently with different materials. For example, relatively low-energy X-rays are much more likely to interact via the photoelectric effect with high-Z materials, and via Compton scattering with low-Z materials. Dose is nice, but how can we account for the different interactions we see with different materials? Specifically, can we know what kind of dose humans get?

Dose equivalent is the answer. We really just need to modify dose so that it depends on the type of radiation. This is done by the **quality factor**, Q, also known as the weighting factor or relative biological effectiveness (RBE). It's a bit like LET. Remember that, in terms of LET, alpha particles deposit the most energy per centimeter, followed by beta, while X- and γ-rays deposit the least. Q weights the dose for damage to biological systems. For γ- and X-rays, and most (>30 keV) beta particles, the value of Q is 1. Beta particles with less than 30 keV have a Q of 1.7. For thermal neutrons (relatively low energy, only 0.025 eV!), it is 5. The Q value varies significantly with the energy of the neutron. The Q value for alpha particles is 20.

We can now define dose equivalent as the amount of biological damage caused by ionizing radiation. Mathematically, it is simply dose (sometimes called "absorbed dose") times Q:

$$H = \text{dose} \times Q \tag{12.1}$$

The SI units for dose equivalent is the Sievert (Sv):

$$Sv = Gy \times Q$$

In the United States, units of rem (*Röntgen equivalent man*) are commonly used.

$$1 \text{ rem} = 10^{-2} \text{ Sv} = 1 \text{ cSv} = 10 \text{ mSv}$$

Since Q is 1 for γ-rays, X-rays, and most beta particles, dose is numerically equal to dose equivalent. For this reason the simpler term "dose" is often used instead of "dose equivalent."

$$1 \text{ Sv} = 1 \text{ Gy} \quad \text{and} \quad 1 \text{ rad} = 1 \text{ rem}$$

We've defined a lot, and it should be enough... but there's another way to look at dose and dose equivalent: **kerma**. It stands for *kinetic energy released in media*. As its name implies, kerma is another way of looking at dose, but instead of energy absorbed by the material, we're looking at energy released by the radiation—a subtle distinction at best. It even has the same units as absorbed dose (Gy). Differences between kerma and absorbed dose are only seen when significant amounts of secondary radiation (bremsstrahlung, secondary electrons) escape the material, thereby getting some of the energy out of the material that was originally transferred to it. Under most circumstances, kerma really ends up numerically equal to dose and exposure.

To relate different media, we can use a simple ratio to adjust dose or kerma to a medium. That ratio is called the *f* **factor** (aka f_{medium} factor) and is similar to the quality factor, Q.

$$f = \frac{\text{dose in medium}}{\text{dose in air}} \tag{12.2}$$

Dose depends mostly on the mass attenuation coefficient (μ_m—Chapter 5), which will depend on the atoms (Z) and the density of the matter. If we are only concerned with humans, we know that water and muscle are made up of atoms with similar atomic numbers (Z). Therefore, we would expect them to have similar *f* factors. Using the *y*-scale on the left side of figure 12.1, and plotting the ratios of the mass attenuation coefficients vs. photon energy, the *f* factor is very close to 1 for most biological materials.

The *f* factors for bones are also plotted on figure 12.1 using the right *y*-scale. Notice that it is also close to 1 for all but the lowest-energy photons. Therefore, the *f* factor is close to one for most biological systems under most conditions, just as Q is close to one for common forms of ionizing radiation.

As is evident from figure 12.1, each of the body's organs will respond a little differently to ionizing radiation. While the differences are subtle, they are sometimes distinguished. **Effective dose equivalent** (sometimes just called **effective dose**) is a

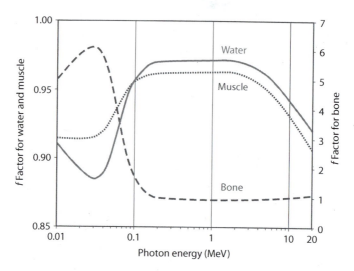

FIGURE 12.1 Variation in the *f* factor for water, muscle, and bone with photon energy. (Data from Hubbell, J. H. and Seltzer, S. M., *Tables of X-ray Mass Attenuation Coefficients and Mass Energy-Absorption Coefficients*. National Institute of Standards and Technology, 2004. http://physics.nist.gov/PhysRefData/XrayMassCoef/cover.html.)

dose equivalent weighted according to susceptibility of a particular organ to ionizing radiation. Going through detailed weighting schemes for different organs is beyond the scope of this text. The reader should simply know that dose can be further refined to a specific organ.

12.2 REGULATIONS AND RECOMMENDATIONS

The **Nuclear Regulatory Commission** (NRC) controls the use and distribution of radioactive materials in the United States. It has primary regulatory authority and is a federal agency. Many individual states have been delegated some of the NRC's responsibilities—Wisconsin is one. In Wisconsin, the state Department of Natural Resources (DNR) handles much of the regulation of radioactive materials.

The NRC *regulates* based on *recommendations* by the International Commission on Radiological Protection (ICRP) and the National Council on Radiation Protection and Measurement (NCRP). Table 12.1 shows NCRP recommendations for doses.

Actual regulations will vary from state to state and will depend on the nature of the facility. For example, the regulations will be much more extensive for a nuclear power plant than for an academic laboratory that handles only very small amounts of radioactive materials.

All doses in table 12.1 are over and above those from natural background sources. The first sets of limits are for radiological workers—notice that the dose for extremities (500 mSv) is much higher than the annual (whole body) dose (50 Sv). Our hands, feet, and skin can take a much greater dose because the cells in these parts of the body can better repair or replace damaged ones.

TABLE 12.1

Effective Dose Limits Recommended by the NCRP

Occupational exposures

Effective dose limits

(a) Annual	50 mSv
(b) Cumulative	10 mSv×age

Equivalent dose limits for tissues and organs (annual)

(a) Lens of eye	150 mSv
(b) Skin, hands, and feet	500 mSv
Guidance for emergency occupational exposure	500 mSv

Public exposures (annual)

Effective dose limit, continuous or frequent exposure	1 mSv
Effective dose limit, infrequent exposure	5 mSv

Effective dose limits for tissues and organs

(a) Lens of eye	15 mSv
(b) Skin, hands, and feet	50 mSv

Education and training exposures (annual)

Effective dose limit	1 mSv

Equivalent dose limit for tissues and organs

(a) Lens of eye	15 mSv
(b) Skin, hands, and feet	50 mSv

Embryo-fetus exposures (monthly)

Equivalent dose limit	0.5 mSv

Source: NCRP Report #116, *Limitation of Exposure to Ionizing Radiation*, 1993.

The "emergency occupational exposure" is designed for life or death situations. If a life can be saved or the spread of significant radioactive contamination be contained, then up to a 500 mSv (whole body) dose can be allowed.

Public exposures refer to nonradiological workers who may be exposed while touring a facility where they may be exposed. It is also exposure due to releases from these facilities. Again, these exposures are over and above those from natural background.

Educational dose limits are designed mostly to allow students to perform experiments with small amounts of radioactive material, or radiation-generating devices. They are also set low because while the body is still growing it is more susceptible to radiation damage. The human fetus is an extreme example, so dose limits are set quite low for pregnant females—to protect the fetus. Note that the human fetus is surrounded by a fair amount of shielding (aka Mom!) and is therefore naturally protected from ionizing radiation.

How do we know what kind of dose we're getting? By measuring exposure! Exposure is measured with TLD (Chapter 6) or film badges, which are worn between the neck and the waist. This gives a pretty good idea of whole body exposure (and dose if we're dealing with photons and/or beta particles). Exposure to the hands is often measured separately using a TLD ring. The ring should be rotated so that the TLD crystals are closest to the source of ionizing radiation. Pocket dosimeters (ionization chambers—Chapter 6) can also be used, especially if the potential for significant short-term exposure is high.

In areas where radioactive materials are used, regular swipe tests are performed to test surfaces for radioactive contamination. Typically a round piece of filter paper is wiped over a random spot on a surface then counted using an organic scintillation counter.

Personnel leaving an area where radioactive materials are handled are required to frisk (monitor) their hands and feet with an appropriate detector (usually a GM tube). If the potential for contamination of other parts of the body exists, those parts must also be monitored. In these situations, a monitoring portal is typically used. Monitoring portals surround all or most of the body with detectors. It was monitoring portals at Swedish and Finnish nuclear power plants that first alerted the world to the Chernobyl disaster. Workers at these plants were *arriving* at work with enough contamination to set them off.

In areas where radiation-generating devices are used, exposure is determined by performing a radiation survey around the equipment while energized. In some cases, TLD or film badges may be placed in certain locations to measure exposure over longer periods of time.

Records are meticulously maintained for all measurements of exposure. As required by law, radiological workers must receive regular (typically annual) reports of their exposure. You should always know what you are getting, and it will likely be well under the limits.

At what doses levels do bad things happen? Data are available from the survivors of the Hiroshima and Nagasaki bombings, the Chernobyl disaster, as well as a number of rather questionable experiments performed by the United States government in the 1950s and early 1960s.[1]

The following lists human health effects at various dose equivalent levels. Keep in mind that these are dose equivalents for a single exposure event. If you are exposed to 25 rem over a period of 10 years, you will not exhibit the symptoms below.

- **25–50 rem** (250–500 mSv): This is when the first measurable biological effects are observed in humans. A decrease in the white blood cell count is seen in a few hours. Later, a measurable decrease in the red blood cell count and in the number of platelets will be observed. Temporary sterility (both genders) is also commonly observed. Complete recovery can be expected in a matter of weeks or in a couple of months. No one dies from this dose.
- **200 rem** (2 Sv): Reversible bone marrow damage occurs. Nausea, vomiting, fatigue, and hair loss also happen. Some will die.

---

[1] Some argue that the U.S. government is continuing to experiment. Veterans of the Persian Gulf wars are currently being monitored for long-term health effects due to exposure to depleted uranium (Chapter 9) used in armor-piercing weapons.

- **400–600 rem** (4–6 Sv): A complete shutdown of the bone marrow function (production of blood cells!) happens, but is only temporary. Severe gastro-intestinal distress is experienced. There's only a 50:50 chance of survival.
- **700 rem** (>7 Sv): The bone marrow function is irreversibly damaged. Death is almost certain.
- **1000 rem** (10 Sv): You will get really, really sick and die. Death will occur within 1–2 weeks.
- **2000 rem** (20 Sv): The nervous system is damaged. Loss of consciousness within minutes; death typically follows in a matter of hours (a day or two at the most).

So what are we exposed to in daily life? Figure 1.6 gives the average dose equivalent to humans living in the United States from a variety of sources. Individual doses will vary. For example, folks living in Denver experience twice as much background radiation as people living at sea level. Note that the units on the y-axis are mrem per year. Radon alone gives us a 200 mrem (or 0.2 rem or 2 mSv) dose every year. The total average dose in the United States is about 380 mrem (0.38 rem) per year. About 79% of that dose comes from natural sources and ~19% is due to medical procedures. Only the remaining 2% is due to humans spreading radioactive materials around the planet.

Not all medical procedures zap you to the same extent. Computed tomography (CT) scans are the worst, resulting in a 3–10 rem dose—depending on how much of you gets scanned. A chest X-ray gives you only ~10 mrem. Nuclear medicine procedures result in doses ranging from 0.1 rem up to 1 rem.

We can compare these procedures to other doses. During an airline flight humans receive ~0.5–1 mrem/h because planes fly up where there's not much atmosphere to shield cosmic rays. Fallout from aboveground nuclear weapons testing is responsible for less than 1 mrem/year of our annual dose. The worst doses received as a result of the Three Mile Island accident in 1979 were ~100 mrem—well below the threshold to cause immediate health effects.

12.3 RISK

Taken collectively, the numbers above should provide us with some reassurance concerning the doses received in medical procedures, or even living through a nuclear power plant accident of the same magnitude as TMI. Even with this safety in numbers, some apprehensions remain. Risk is something that can be quantified and understood, but for most of us, it is our perception of risk that affects our behavior.

Table 12.2 gives the relative risk of various activities and technologies. Starting from a list of 30, a group of 97 college students in two Radiation Physics classes were asked to quickly rank them from 1 to 30, with 1 being the riskiest and 30 being the safest. Their collective results appear under "Class." Only the top 10 and selected others are shown. These results can be considered "perceived" risk.

Table 12.2 also gives rankings of some of the same activities and technologies by professionals that study risk ("Experts"). These rankings can be considered "actual" risk. By and large, the collective intelligence of the class correctly identified most

TABLE 12.2

Rankings of Risky Activities and Technologies

Class	Experts
1. Smoking	1. Motor vehicles
2. Motorcycles	2. Smoking
3. Handguns	3. Alcoholic beverages
4. Motor vehicles	4. Handguns
5. Firefighting	5. Surgery
6. Surgery	6. Motorcycles
7. Police work	7. X-rays
8. Alcoholic beverages	8. Pesticides
9. Large construction	9. Electric power (nonnuclear)
10. Hunting	10. Swimming
12. Pesticides	13. Large construction
14. X-rays	17. Police work
15. Nuclear power	18. Firefighting
17. Electric power (nonnuclear)	20. Nuclear power
29. Swimming	23. Hunting

Source: "Expert" rankings, Slovic, P., *Radiat. Prot. Dosimetry*, 68,
165, 1996. With permission.

of the top 10 riskiest things. There are, however, some fascinating anomalies in the overall rankings by the two groups. These anomalies are discussed in the following paragraphs.

Familiarity breeds contempt. The class ranked swimming as much safer than the experts, likely because it is something most of us have done safely, and likely with some pleasure. The stats say otherwise. Lots of people drown every year while swimming, making it an inherently dangerous activity. The students also ranked X-rays quite a bit lower than the pros. Again, this is because many of us have had X-rays, and suffered no ill effects. However, X-rays are a form of ionizing radiation and are dangerous to biological systems like us. A similar argument could be made for why motor vehicles appear somewhat lower on the class list than on the experts'.

Lack of control creates fear. Most people would rank nuclear power much higher than the experts and, despite being in a radiation physics class, the students followed suit. With the exception of X-rays, anything involving ionizing radiation creates an unusual amount of fear in the general population. As a result, even groups with some higher education will list nuclear power as more dangerous than is borne out by experience. Ionizing radiation is feared, in part, because we cannot directly sense it, and we know that it *can* cause adverse health effects, sometimes decades in the future. Our lack of control over exposure to and effects of ionizing radiation causes it to be seen as a higher risk than is warranted. Additionally, nuclear power is feared because of its potential for catastrophe—a nuke plant burning is much more dangerous to the surrounding public than a coal plant catching fire.

Training and experience make certain activities safer than we think. The students ranked police work, firefighting, large construction, motorcycles, and hunting as more dangerous than they really are. While there is real risk associated with these activities, they are less dangerous than we think because of the extensive training, detailed procedures, and considerable regulation associated with each. The same is true of radiation therapy and nuclear medicine. Because both professions involve work with ionizing radiation, and because we (as a society) fear radiation, they are highly regulated. Only well-trained professionals are allowed to perform these procedures, and they do so with a remarkable safety record. These professions are among the safest of all careers! The same can be said of nuclear power. Because it is more extensively regulated than many other forms of power generation, it is inherently safer.

We've discussed risk in general terms and have tried to show that there is often a difference in the perception of risk and the actual risk. We've also looked at the health effects of receiving a single large dose. Those of us working with radioactive materials and radiation-generating devices would much rather know: what are the health risks at low doses?

There is still some debate about the answer to this question. Because the odds of adverse effects become so small at low doses, it is extraordinarily difficult to quantify them. The most accepted model is that there is a linear relationship between risk and dose. However, the fact that so many people have lived in areas (such as Denver) where the natural background is much higher and not suffered any measurable ill effects suggests that low radiation doses may not be harmful. As suggested in figure 12.2, the risk may drop faster than the linear projection suggests. It may also rise above linearity. Because of the uncertainty of the consequences of low doses, it is still important to minimize dose, even if it appears that a higher dose poses no greater risk.

All radiological workers face the risk of radiation exposure—how do we manage it? As mentioned earlier, we monitor it closely through badges, rings, swipe tests, and radiation surveys. Most importantly when working with radiation, we practice **ALARA**. As mentioned in Chapter 1, ALARA stands for "*as low as reasonably achievable.*" We want to apply this concept to ourselves, those we work with, and to the environment in terms of contamination. Just because safety limits are set, doesn't mean that we have to run our doses up to them. For example, it there is a procedure that can be done more than one way with equal effectiveness or benefit, we should pick the procedure that results in the lowest dose and contamination.

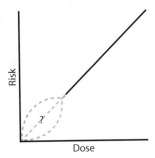

FIGURE 12.2   Risk vs. dose.

Exposure can be lowered by minimizing **time** near the source, maximizing **distance** to the source, and increasing **shielding** around the source. Exposure or dose will change in a linear fashion with time. If time near a source is doubled, dose will be doubled.

Example: If the dose rate near a source is 6.6 mSv/h, what dose will be received over 30 min?

$$\frac{6.6 \text{ mSv}}{h} \times 0.50 \text{ h} = 3.3 \text{ mSv}$$

Exposure or dose varies inversely with the square of the distance. If distance is doubled, exposure is decreased by a factor of four. If distance is tripled, exposure decreases by a factor of 9. This effect is known as the **inverse square law**.

Example: If the dose rate 1.0 m from a source is 6.6 mSv/h, what will the dose rate be 4.0 m from the source?

$$\frac{6.6 \text{ mSv}}{h} \times \frac{1}{d^2} = \frac{6.6 \text{ mSv}}{h} \times \frac{1}{\left(4.0 \text{ m}/1.0 \text{ m}\right)^2} = \frac{0.41 \text{ mSv}}{h}$$

We can combine time and distance into a single equation:

$$X = \frac{\Gamma \cdot A \cdot t}{d^2} \tag{12.3}$$

where X is exposure in Röntgens (R), and Γ is the exposure rate constant, which *can* have units of $R \cdot cm^2/mCi \cdot h$. The same equation can be used for dose or dose equivalent, so long as the units are changed appropriately. If the units of the exposure rate constant are as above, then A is activity (mCi), t is time (h), and d is distance (cm). Activity, time, and distance units can also be changed (e.g., Ci instead of mCi), but should be consistent with the units on the exposure rate constant.

Example: A human sits 1.5 m from 25 mCi of ^{137}Cs for 40 h each week. What is her exposure? The exposure rate constant is $3.27 \text{ R} \cdot cm^2/mCi \cdot h$.

$$X = \frac{3.27 \text{ R} \cdot cm^2 / mCi \cdot h \times 25 \text{ mCi} \times 40 \text{ h}}{(150 \text{ cm})^2} = 0.145 \text{ R} = 145 \text{ mR}$$

Shielding is a little less straightforward. Appropriate shielding should be added when necessary. Low-Z materials (like plastics) should be used to shield pure β^--emitting nuclides. If there's a fair amount of the β^--emitting nuclide, significant bremsstrahlung photons may be produced from the interactions of beta particles with higher Z materials. The use of a low-Z material (like plastics) for shielding will therefore minimize the amount of bremsstrahlung. If bremsstrahlung is unavoidable, or the nuclide emits γ-rays as part of its beta decay, a high-Z, high-density material like lead should surround the plastic shielding.

High-Z, high-density materials are the best choice for shielding energetic photons such as X-rays and γ-rays. Remember that the probability of photon interaction increases with the atomic number (Z) and the amount of material it passes through (density). Also recall that half value layer (HVL) is the thickness of material that will attenuate half of the photon beam passing through it and that TVL is the tenth value layer—the thickness of material that will cut a photon beam to one tenth of its original intensity. If two TVLs are used as shielding, the photon beam will be cut to 1/100 of its original intensity.

Example: How much lead is needed to cut the exposure in the example above to 2 mR/week?

^{137}Cs is a β$^-$ and γ emitter. Let's just deal with the γ-rays. If μ = 1.07 cm^{-1} for Pb exposed to γ-rays from ^{137}Cs, then:

$$\ln \frac{I_0}{I} = \mu x$$

$$\ln \frac{145 \text{ mR}}{2 \text{ mR}} = 1.07 \text{ cm}^{-1} x$$

$$x = 4.0 \text{ cm}$$

How many HVLs is this?

$$HVL = \frac{\ln 2}{\mu} = \frac{\ln 2}{1.07 \text{ cm}^{-1}} = 0.65 \text{ cm}$$

$$\frac{4.0 \text{ cm}}{0.65 \text{ cm}} = 6.2 \text{ HVLs}$$

How many TVLs is this?

$$\ln \frac{10}{1} = \mu \times TVL$$

$$TVL = \frac{\ln 10}{\mu} = \frac{\ln 10}{1.07 \text{ cm}^{-1}} = 2.2 \text{ cm}$$

$$\frac{4.0 \text{ cm}}{2.2 \text{ cm}} = 1.8 \text{ TVLs}$$

We can incorporate shielding in the exposure equation given earlier:

$$X = \frac{\Gamma \cdot A \cdot t}{d^2} B \tag{12.4}$$

where B is the **transmission factor**—the fraction of radiation making it through the shielding. B has no units and is equal to I/I_0.

Example: A human works 40 h/wk 1.0 m from a 100 mg ^{226}Ra source. If the exposure rate constant is 0.825 R·m²/h·Ci (note change in units!), what thickness of lead would lower the weekly exposure to 0.10 R? $\mu = 0.46$ cm^{-1}

First, we need to calculate the activity of the source in Ci:

$$A = kN = \frac{\ln 2}{1599\ a \times \left(5.26 \times 10^5\ \text{min/a}\right)} \times \left(100\ g \times \frac{\text{mol}}{226.0\ g} \times \frac{6.022 \times 10^{23}\ \text{atoms}}{\text{mol}}\right)$$

$$= 2.20 \times 10^{11}\ \text{dpm}$$

$$2.20 \times 10^{11}\ \text{dpm} \times \frac{\text{Ci}}{2.22 \times 10^{12}\ \text{dpm}} = 0.0990\ \text{Ci}$$

Now we can solve the exposure equation for B:

$$B = \frac{X \cdot d^2}{\Gamma \cdot A \cdot t} = \frac{0.10\ \text{R} \times (1.0\ \text{m})^2}{\left(0.825\ \text{R} \cdot \text{m}^2/\text{h} \cdot \text{Ci}\right) \times 0.0990\ \text{Ci} \times 40\ \text{h}} = 0.031$$

Since $B = I/I_0$:

$$\ln\frac{I}{I_0} = -\mu x \quad \ln 0.031 = -0.46\ \text{cm}^{-1} \times x \quad x = 7.6\ \text{cm}$$

Let's look at one more problem, but this time let's look at dose equivalent rate (DR) rather than exposure. Keep in mind that dose equivalent rate is dose equivalent/t. Now Γ is the dose equivalent rate constant.

$$\frac{\text{dose equivalent}}{t} = DR = \frac{\Gamma \cdot A}{d^2} \tag{12.5}$$

Example: Calculate the dose equivalent rates at 10 and 300 cm from a syringe containing 27 mCi of ^{99m}Tc. The dose equivalent rate constant for ^{99m}Tc is 3.32×10^{-5} mSv·m²/MBq·h.

First, we'd better convert activity to MBq.

$$A = 27\ \text{mCi} \times \frac{\text{Ci}}{1000\ \text{mCi}} \times \frac{3.7 \times 10^{10}\ \text{dps}}{\text{Ci}} \times \frac{1\ \text{Bq}}{1\ \text{dps}} \times \frac{\text{MBq}}{10^6\ \text{Bq}} = 1000\ \text{MBq}$$

At 10 cm$=0.10$ m:

$$DR = \frac{3.32 \times 10^{-5} \text{ mSv} \cdot \text{m}^2 / \text{MBq} \cdot \text{h} \times 1000 \text{ MBq}}{(0.10 \text{ m})^2} = 3.3 \text{ mSv/h}$$

At 300 cm$=3.0$ m:

$$DR = \frac{3.32 \times 10^{-5} \text{ mSv} \cdot \text{m}^2 / \text{MBq} \cdot \text{h} \times 1000 \text{ MBq}}{(3.0 \text{ m})^2} = 0.0037 \text{ mSv/h} = 3.7 \text{ } \mu\text{Sv/h}$$

QUESTIONS

1. Calculate the number of ion pairs formed in 1.00 cm^3 of air ($\rho = 1.29$ g/L) exposed to 1.00 R of X-rays.
2. Assuming that all of the X-ray energy in the previous problem is deposited in the air, and that the average ionization energy of air can also be expressed as 33.85 J/C, what dose is deposited? Calculate your answer in both Gy and rad.
3. The average American receives a 200 mrem dose from radon each year. Assuming you receive the average dose, and that it all comes from the 5.49 MeV alpha particle emitted by ^{222}Rn, how much energy is deposited in your body each year from radon? Approximately how many decays does this represent?
4. How much energy (MeV) is deposited in a 75 kg human from a 55 mrad dose? What is the maximum number of ion pairs that could be formed in air from this same dose?
5. Calculate the dose equivalent in humans, in both mrem and Sv, for the dose in the previous problem of alpha radiation. Do the same for 55 mrad of X-radiation.
6. During a routine R–T procedure, a therapist is exposed to 310 mR/h. If the procedure takes 4.0 min, what is her exposure? These data were obtained while she was standing 0.91 m from the source. If she stands 0.61 m from the source, she can complete the procedure in 1.5 min. Would her exposure change? If so, by how much?
7. How does ALARA apply to a patient about to undergo a radiological procedure?
8. Using figure 1.6, estimate the average total dose for a U.S. citizen. What percentage of this dose is anthropogenic? Compare to the annual public effective dose limit established by the NRCP. Briefly explain any discrepancy.
9. A person works 24 ft from a 1.14 mg (tiny!) sample of ^{60}Co. Assuming a 40 h work week with two weeks off per year, and no shielding, what would her annual dose be (mSv)? The TVL in lead for the γ-rays emitted by ^{60}Co is 36.2 mm. Calculate the thickness of lead needed for a 10 mSv annual dose. Briefly state two other ways exposure could be decreased.

10. A patient is apprehensive about receiving the 15 mrem dose you are about to administer. What should you do?

11. A patient receives a 15.4 µCi of ^{131}I as part of a medical procedure, then goes home to sleep. What maximum dose equivalent could the patient's wife receive overnight if they sleep 48 cm apart for 8 h and 20 min? Assume that the gamma emissions from the ^{131}I travel an average of 15 cm before leaving the patient's body. The linear attenuation coefficient for ^{131}I in humans is 0.11 cm^{-1}, and the dose equivalent constant for ^{131}I is 7.47×10^{-5} mSv·m^2/MBq·h.

12. The exposure 182 cm from a source is 21 mR over 5.0 min. What is the exposure 122 cm from the same source for 1.5 min?

13. The dose 0.91 m from a source is 2.6 mSv over 4.0 h. What is the dose 0.33 m from the same source for 1.7 h?

13 X-ray Production

X-rays are used extensively in diagnostic and therapeutic medical applications. This chapter starts by looking at how X-rays are generated for diagnostic applications such as those performed in dentist and doctor offices, and then moves into how high-energy X-ray beams are produced for radiation therapy. Interestingly enough, knowledge of how electrons interact with matter (Chapter 5) is key to understanding how X-rays come about.

13.1 CONVENTIONAL X-RAY BEAMS

For diagnostic medical applications, such as an X-ray of your foot to see if any bones are broken, we'd like to be able to consistently produce high-energy photons on demand—a flip of the switch to provide the same energy and intensity every time would be great. It is also desirable for our photon generator to be safe and affordable. The only source that satisfies all these requirements is a particle accelerator. A **conventional X-ray tube** is a small particle accelerator (Chapter 7). Let's see how it works.

A schematic of a typical conventional X-ray tube is shown in figure 13.1. The filament is a tungsten wire, much like those used in incandescent light bulbs. If we run a little electric current through the filament at a high enough voltage, we can get electrons to boil off the wire. The filament is the cathode.

After they jump off the cathode, the electrons are accelerated toward the positively charged anode (target). The maximum energy they attain will be equal to the potential difference (tube voltage) between the two electrodes. For all this to happen, both electrodes need to be under vacuum, so the whole X-ray tube is enclosed, and pumped out. This is important because the electrodes would rapidly corrode if exposed to oxygen, and because the electrons could interact with any matter (gases) on their way from the filament to the target.

Electrons won't boil off the cathode unless the filament is pretty hot. The current of electrons flowing through the wire (**filament current**) does this—just like a toaster! Since it will get pretty hot, it's important to make the filament out of a metal with a high melting point. Tungsten melts higher than any other metal (3370°C!!), so it works quite well. The filament current is usually set at a few amps. The other current flowing in the tube is made up of the (relatively few) electrons moving from the cathode (filament) to the anode (target). This is called the **tube current** and is usually a few tenths of an amp (few hundred milliamps).

The **focal spot** is the place on the target where the electrons hit the target. A smaller focal spot creates a smaller beam, which is sometimes desirable. For example, minimizing the dose to surrounding tissue when trying to kill a small tumor. A smaller focal spot can be accomplished by using a smaller filament (makes for a smaller electron beam). Unfortunately, this also creates a weaker (less intense) X-ray beam.

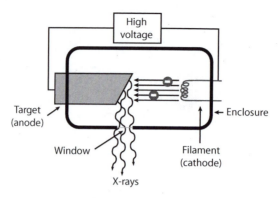

FIGURE 13.1   Schematic of a conventional X-ray tube.

So, how does an X-ray tube make X-rays? When the electrons hit the target, there's two ways they can make X-rays (see Chapter 5 for more details):

1. Bremsstrahlung—the continuous energy spectrum of X-rays produced through inelastic scattering of the electrons in the tube current by the nuclei in the target. This can always happen, regardless of the energy of the incoming electrons. The maximum X-ray energy would be equal to the energy of the electron (tube voltage).
2. Characteristic X-rays—if the beam electrons have enough energy to excite or ionize a K-shell electron in the target material, an electron vacancy will be created in that shell. When an L- or M-shell electron drops down to the K shell to fill the vacancy, a characteristic X-ray is emitted. Remember, a characteristic X-ray has a specific energy, because the target electron energy levels are quantized.

When we put these two types of interactions together, we see an interesting distribution of photon energies—as shown by the solid line in figure 13.2. The rather broad part of this spectrum (almost all of it!) is bremsstrahlung radiation. Notice that it maxes out at fairly low energy then gradually tails off as energy increases until just below the tube voltage. The probability of forming lower-energy photons via bremsstrahlung is higher, because it is more likely that an electron will interact with an atom at a longer rather than at a shorter distance. The bremsstrahlung feature tails off sharply at low energy because the very-low-energy X-rays are more efficiently filtered by the tube enclosure. The dashed line in figure 13.2 shows what the bremsstrahlung spectrum would look like without this filtration.

The sharp peaks on top of the bremsstrahlung spectrum are the characteristic X-rays produced from ionizations and excitations. The energies of these peaks depend on the target material. Tungsten is used as a target for medical applications because it is high melting, and because its characteristic X-rays are relatively high in energy. The energy increases because the spacing between electron orbitals generally increases with atomic number. Higher energy means more penetrating—less likely to interact with matter. If only low-energy (soft) X-rays were used, they would be almost entirely

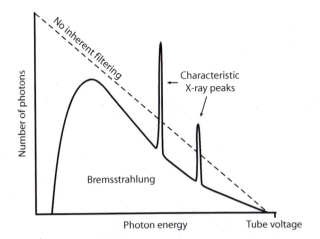

FIGURE 13.2 X-ray spectra with (solid line) and without (dashed line) inherent filtering.

absorbed by the patient. A higher atomic number for the target also means that the intensity of the X-rays produced via bremsstrahlung will be higher. Remember that as Z increases, the more likely it is for an electron to interact (Chapter 5).

X-rays can be produced at different depths inside the target—not just at the surface. If one is produced a little ways in, it could interact (scattered or absorbed) with the target material on its way out.

We've already seen that the X-ray tube enclosure filters the X-rays to some extent. Because the heat produced in the process of making X-rays needs to be dissipated, X-ray tubes are usually surrounded by oil or water that is circulated. Finally, there is usually an exit window for the X-rays to travel through before leaving the tube. The total attenuation due to the enclosure, coolant, and exit window is referred to as the tube's **inherent filtration**. There's nothing we can do about this; it is a result of the tube's design.

The inherent filtration is usually equivalent to placing ~1 mm of Al in the X-ray beam. Like all forms of photon filtration, inherent filtration will harden the X-ray beam. Remember that this means the overall intensity is lower, and the average energy is higher after filtration (Chapter 5). Additional hardening can be performed after the beam exits the tube by placing additional filters in the beam's path. This is desirable for most medical applications since the low-energy photons will tend to deposit all of their energy in the first centimeter or so of the matter (flesh!). Although the average energy of a polyenergetic photon beam is increased with greater filtration, the overall intensity of the beam will be lower.

What happens to the X-ray spectrum when we mess with the current? Basically, it just changes the overall intensity. Intensity increases with current. Increasing tube current increases filament current—more electrons bombard the target, creating more X-rays.

How about voltage? When voltage is increased, the average energy of the photons produced is also increased. Sometimes the low-energy X-rays are desirable. If we wanted to treat a surface malady, such as skin melanoma, then we'd want as many low-energy photons as possible. Surface treatments are usually performed with 5–15 keV photons. If we want to go a little deeper (~5 mm below the surface), then 50–150 keV would be appropriate. If we need to penetrate further, we need to keep cranking up the voltage.

The shape of the target is also important to consider. In figure 13.1, the electron beam hits the anode at an angle. This angle serves two purposes. First electrons with less than 1 MeV of energy (true for all conventional tubes) *tend* to produce X-rays in a direction that is 90° from the electron beam. This process is called "**reflectance**" because it *looks* like light reflecting on a mirror. You should know this is not true!

Second, a focal spot on an angle to the electron beam will produce a narrower, more intense X-ray beam. The angle allows a wider electron beam to hit it (higher intensity X-rays produced), while producing an X-ray beam that is smaller in area—as pictured in figure 13.3. The length A of the focal spot (width of the electron beam) is larger than the length a (width of the X-ray beam). In fact, the mathematical relationship between the two lengths is:

$$a = A \times \sin \theta \tag{13.1}$$

where the angle θ is the angle of the anode. In figure 13.3, this angle is 30°. If A is 0.10 cm, then:

$$a = 0.10 \text{ cm} \times \sin 30° = 0.050 \text{ cm}$$

As mentioned above, reflectance only happens with electron energies below 1 MeV. When electrons have more than 1 MeV, X-rays *tend* to be produced in the same direction as the electron beam. This process is called "**transmission**" because it appears that electrons are transmitted though the target, turning into X-rays (fig. 13.4). As you might be able to guess, the process of "converting" an electron beam to an X-ray beam is a bit more complex than suggested by figure 13.4!

Keep in mind that X-rays only *tend* to be traveling in the directions shown in figures 13.3 and 13.4. In reality, X-rays will travel out of the target in lots of different directions, regardless of how they are made. Since we are interested in generating a *beam* of X-rays all traveling in the same direction, we want to only select photons traveling in the same direction, and soak up all the others. For this reason, X-ray tube housings are usually pretty thick and are made up of high-Z metals. To deselect the photons that are traveling close to the correct direction, they are passed through a metal collimator, as shown in figure 13.5. The length and aperture of the

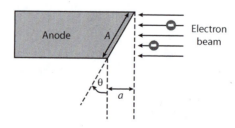

FIGURE 13.3 Geometric reduction of the beam size relative to the focal spot size.

FIGURE 13.4   A transmission target.

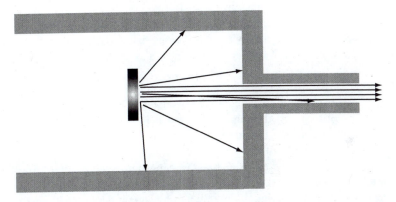

FIGURE 13.5   The use of a collimator to form an X-ray beam.

collimator will determine the consistency (how parallel are the photons?) and size of the beam.

Only part of the projectile electrons' kinetic energy goes into X-ray production. Remember that the maximum electron energy in kiloelectronvolts is equal to the tube's electric potential in kilovolts. We can then get a handle on efficiency using:

$$f = 3.5 \times 10^{-4} ZE \qquad (13.2)$$

where f is the fraction of electron energy that is turned into photon energy, Z is the atomic number of the target material, and E is the maximum electron energy in million electron volts. For example, with a tungsten anode and a tube potential of 140 kV:

$$f = 3.5 \times 10^{-4} \times 74 \times 0.140 \text{ MV} = 0.0036 \text{ or } 0.36\%$$

Only 0.36% of the energy being deposited by the electron beam into the anode ends up as X-ray photons! This is so inefficient it makes SUVs look pretty green. The rest of the electrons' energy ends up as heat. This is the source of all the heat that needs to be dissipated by the coolant mentioned earlier. If we repeat this calculation for a 4 MV electron beam, we'll get 10.4% efficiency, and 25 MV gives 65%. As the tube potential increases, so does the efficiency—in rather dramatic fashion. If more photons are desirable, then cranking up the voltage is one way to get 'em.

13.2 HIGH-ENERGY X-RAY BEAMS

As described in Chapter 11, the history of the use of X-rays for radiation therapy shows ever-increasing photon energy. Conventional tubes were commonly used in the first half of the twentieth century. Increasing photon energy means more efficient X-ray production (higher intensity), and allows for treatment deeper inside the patient.

Instruments using conventional tubes have a number of names, usually related to the energy of the X-rays produced. Superficial therapy machines operate in the 50–150 kV range, and were only useful in treating the surface or near surface of the patient's body. **Orthovoltage** machines were in common use in the middle of the last century, but are limited to about 400 kV. They were also known as "deep therapy" machines, because of their ability to treat lesions below the skin. They have now been completely replaced by higher-energy linear accelerators, the so-called **megavoltage** machines. Why?

Reason #1: Lower-energy photons are more likely to interact with matter. Basically, they are not penetrating enough—too much energy is dumped in the first centimeter (or so) of tissue. If the tumor is deeper than that, we'll end up giving the healthy tissue between the skin and the tumor too much of a dose. To deliver more of a dose deeper into the patient, we need more power!

We already know this is an issue, but how can we imagine this in terms of dose delivered to humans? This is pictured in figure 13.6. Percentage depth dose (PDD) is the percentage of the maximum dose remaining at a particular depth in a human. If the maximum dose is delivered 1 cm below the surface, and the dose at 2 cm is 80% of the maximum, then the PDD at 2 cm is 80. Figure 13.6 shows how PDD drops off as the high-energy photons travel through matter. This is exactly what we expect—the thicker the matter, the greater the beam attenuation (Chapter 5).

Several different curves are represented in figure 13.6. In going from the orthovoltage curve to the 18 MV curve, the average energy of the photon beam increases, and so does the degree of penetration—greater doses are delivered to greater depth.

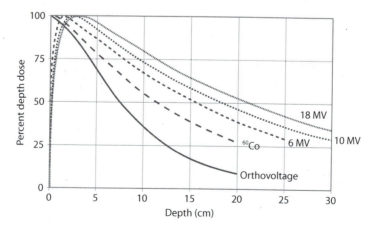

FIGURE 13.6   Variation of percent depth dose (PDD) with depth in water and energy of the X-ray beam.

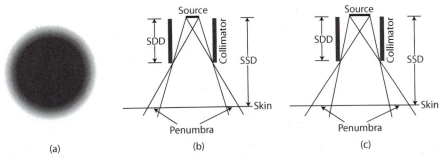

FIGURE 13.7 (a) Penumbra viewed around a cylindrical beam. (b) Geometric origin of penumbra. (c) Effect on penumbra of increasing source size.

These data were collected by placing a detector the appropriate distance underwater while exposing the water to the X-ray beam. Water is an excellent surrogate for humans since we are composed of mostly water.

The ^{60}Co curve lies just above the orthovoltage curve. ^{60}Co emits photons with energies of 1.2 and 1.3 MeV, and therefore represents a higher-energy beam than an orthovoltage machine. The remaining three curves are from megavoltage machines of increasing energies. Notice that the percentage depth dose starts at 100% at the surface of the water for the orthovoltage and ^{60}Co curves. It is not until we get to the higher-energy machines that the peak moves below the surface.

Reason #2: **Penumbra** size (or beam edge unsharpness) is larger for lower-energy beams. Penumbra is the fuzzy edge around the beam. This is true for any beam of photons traveling through matter (fig. 13.7a). Even laser light has a fuzzy edge to it. Penumbra is bad for radiation therapy because the tumor needs to be treated with a uniform dose, and penumbra creates nonuniformity at the beam edge.

Some of the penumbra is due to scattering of photons as they travel through matter (usually air); the rest is due to the size of the source. As can be seen in figure 13.7b, a certain area of the skin will be able to see the entire source, and will receive the maximum dose. As we begin to move outside of that area, the collimator blocks parts of the source, so that area will receive a lower dose. Eventually we move too far away, and the source is completely blocked by the collimator. This area sees no dose. Increasing the size of the source (focal spot in X-ray machines) will increase the size of the penumbra (fig. 13.7b and c).

Decreasing the source-to-collimator distance (SDD) also increases penumbra. Finally, increasing source-to-surface distance (SSD) increases penumbra. Decreasing the collimator–patient distance decreases penumbra, but a certain amount of distance (>15 cm) is a good idea. Photons hitting the collimator can kick out electrons—if the collimator is too close to the patient, they can get a pretty good skin dose from these electrons. Having some distance also facilitates movement of the source around the patient.

Reason #3: Can't use **isocentric techniques**. Rotating the source around the patient with the target at the center of the circle will maximize the dose to the tumor while minimizing the dose to the healthy tissue surrounding the tumor. The point (or volume) of the beams' convergence is the **isocenter**. Figure 13.8 shows the effect of using four different source positions on the dose received—the darker the shading,

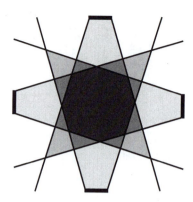

FIGURE 13.8 The effect of using isocentric techniques to maximize dose in the area of convergence, while minimizing elsewhere.

the greater the dose. As you might be able to imagine, using more source positions will further maximize dose to the area of convergence (the tumor) while minimizing it elsewhere (healthy tissue).

Why can't orthovoltage machines use isocentric techniques? Remember that X-rays are very inefficiently produced at orthovoltage energies. Therefore the intensities of these beams are also low—in order to get a decent dose, orthovoltage machines need to be close to the patient ($\sim$50 cm), and cannot be safely rotated around them. Efficiency and intensity increase dramatically at the voltages used by megavoltage machines, allowing them to be far enough ($\sim$100 cm) from the patient to be rotated. An added bonus to moving the source further from the patient is that the extra air the beam now travels through will act as a filter and harden it.

As mentioned in "Reason #1" above and clearly demonstrated in figure 13.6, higher-voltage beams deliver their maximum dose below the surface of the patient (skin). As you remember from Chapter 11, frying the patient's skin was used as an indication they had been zapped enough back in the early days of radiation therapy. Delivering the maximum dose below the surface of the skin is known as **skin sparing**. This phenomenon is simply illustrated in figure 13.9, but it is a little more complex than simply a lower probability of interacting with matter.

Higher-energy photons interact with low-Z matter almost exclusively via Compton scattering (Chapter 5). When this happens, both the scattered photon and the Compton electron tend to move in the same general direction as the incident beam. Because of their higher probability of interacting with matter, the Compton electrons will deposit their energy fairly close to where they are generated, but it will be deeper into the body. Patient dose at depth is therefore due to both photon *and* electron interactions within their body.

Higher-energy beams will be more penetrating, and will transfer more energy to the electrons they knock loose. Higher-energy electrons will travel further into the patient's body before they deposit much of their energy (under the Bragg peak, Chapter 5). As a result the depth of the maximum dose increases with the energy of the incident beam. Table 13.1 illustrates this trend.

It turns out that the Compton electrons are key to figuring out where the maximum dose is delivered. The maximum dose is deposited at the point inside the body

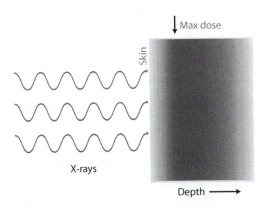

FIGURE 13.9   Skin sparing by higher-energy X-ray beams.

TABLE 13.1
Maximum Dose Depth in Water for Various Photon Energies

Photon energy	Depth of max dose (cm)
Orthovoltage	0.0
^{137}Cs (0.7 MeV)	0.1
^{60}Co (1.2 MeV)	0.5
4 MV	1.0
10 MV	2.5
25 MV	5.0

where the number of electrons knocked loose is equal to the number of electrons being stopped. This is called **electron equilibrium**. Before this point more electrons are being knocked loose and after it fewer electrons are stopping.

At this point you should be pretty well sold on the fact that X-ray machines operating in the MV range will be better for treating most tumors than X-rays generated by kV machines. Unfortunately, conventional tubes cannot accelerate electrons into the MeV range. We need a more powerful particle accelerator. A quick glance back to Chapter 7 reminds us that we can use linear accelerators to accomplish this.

As we saw in Chapter 7, most modern-day linear accelerators manipulate the phase (peaks and troughs) of the microwaves to accelerate electrons. The microwaves can be generated using the same technology as a microwave oven. **Magnetrons** are used in microwave ovens and 4–10 MV accelerators. **Klystrons** are used in higher-voltage machines. Both use electrons to create microwaves via resonance. It works very much the same as blowing air across a bottle top, or air being blown through a pipe organ. Some of the energy of the moving air gets converted into sound through resonance. The same thing happens inside a magnetron or a klystron. Instead of air, it is electrons moving near a cavity, and microwaves are created instead of sound.

Once generated, we want to make sure our waves don't lose any energy. When a broadcast antenna emits radio waves, they spread out in all directions at once.

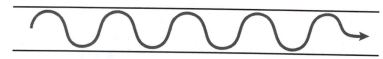

FIGURE 13.10 An electromagnetic wave moving through a waveguide.

FIGURE 13.11 A standing wave accelerator guide.

The farther away from the antenna, the lower the intensity of the waves. In a linear accelerator the microwaves are moved from one point to another with no loss of intensity, by **waveguides**. These can be as simple as copper tubes, but can be more complex (fig. 13.10). The waveguide does just what its name implies, it guides the wave along a specific path, rather than allowing it to spread out in all directions at once. Thereby, the same intensity of waves is seen at the end as is put in at the beginning.

Remember that electromagnetic radiation behaves a lot like waves in water. We can think of phase in relation to the peaks and troughs of a wave. A peak represents one phase and the trough represents the opposite phase. The electron is pushed and pulled by these phases, so if we can arrange them just right, they will accelerate the electron. As also mentioned in Chapter 7, there's another way to look at the acceleration of charged particles inside a waveguide. As the EM wave moves through the guide, it will induce temporary positive and negative electrical charges on the inside surface of the guide. If these temporary charges move just right through the waveguide, they can be used to accelerate charged particles.

Figure 13.10 shows a wave moving from left-to-right through a cylindrical waveguide. This can be used simply to transfer the wave's energy from one point to another, or it can be used to accelerate a charged particle, like an electron. When used to accelerate, it is called a **traveling wave accelerator guide**. We can also use two waves traveling in opposite directions through the same waveguide to set up an interference pattern as illustrated in figure 13.11. This is called a **standing wave accelerator guide**. If the interference is done just right, electrons will be accelerated as they move through the guide.

Remember that *constructive interference* happens when two waves are "in phase"—e.g., the peak of one wave is at the same point as the peak of the other wave. At this point the peak height would be the sum of the two peaks. *Destructive interference* happens when the waves are "out of phase"—the peak of one wave coincides with the trough of the other. In this case, the two would cancel each other out. When two waves interfere as shown in figure 13.11, a complex pattern results. Our concern is not with how to get this just right to accelerate an electron to specific velocity, but rather to know that this is how it is done.

Conventional X-ray tubes can only accelerate an electron to a velocity of approximately one half that of the speed of light (0.5c), and cyclotrons can't do much better.

Linear accelerators are needed to get 'em going faster than that. Keep in mind that particles cannot travel at, or faster than, the speed of light, and as we accelerate a particle to speeds approaching that of light, increasing amounts of energy are required. Using the formulas from Chapter 5, we can calculate the relativistic kinetic energy for an electron at various velocities (table 13.2). We can see that as the velocity approaches the speed of light ($v/c \rightarrow 1$), the amount of energy goes through the roof! This means that we can get very-high-energy electrons if we can accelerate them to high enough velocities.

Where do we get our electrons for linear accelerators? The same way they are produced for conventional tubes—boil them off a hot, high-voltage filament under vacuum. This time we drill a hole in the anode right where the focal spot would be so the accelerated electron passes right through (fig. 13.12) into the accelerator guide. This is the same technology used in cathode ray tubes (CRT—old-style televisions and computer monitors). When placed at the end of a linear accelerator, it is called an **electron gun**. Mucho macho.

We can now put all of our components together. Figure 13.13 shows a schematic for a megavoltage X-ray machine that uses a magnetron microwave generator. The microwaves travel through the waveguide to the accelerator guide. There, they accelerate the electrons emerging from the electron gun and slam them into the target. A transmission target is used for all megavoltage machines. Its thickness is a careful balance between the need to stop all the electrons (thicker is better) and the desire to avoid absorbing the photons that are generated (thinner is better). Targets for higher-energy machines will be thicker than those operating at lower energies.

Notice how the accelerator guide fits vertically in the machine's head in figure 13.13. For higher-energy accelerators, a longer accelerator guide is needed, so

---

TABLE 13.2
Relativistic Kinetic Energies for an Electron at Velocities Approaching the Speed of Light

v/c	KE (MeV)
0.50	0.079
0.90	0.66
0.990	3.11
0.9990	10.92
0.9999	35.63

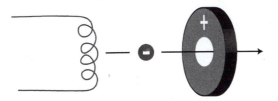

FIGURE 13.12   A cathode ray tube, aka an electron gun.

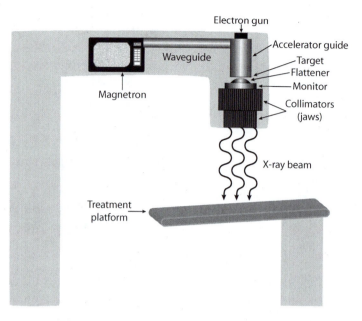

FIGURE 13.13   A MV X-ray machine with a magnetron and a vertical accelerator guide.

it is placed horizontally, as pictured in figure 13.14. Both types of machines use a thin ionization chamber (Chapter 6) as a **monitor** to measure the radiation flux, and thereby determine the patient's dose. An ionization chamber is used precisely because it is very inefficient in detecting high-energy photons. We want to get an idea of the flux without significantly attenuating the beam. Both figures 13.13 and 13.14 have collimators (aka jaws) to help shape the beam. They are typically made from high-Z materials (like Pb). Finally, there's a treatment platform (slab!) for the patient to lie on.

In higher-energy machines (fig. 13.14), the electron beam needs to be bent by 90° so it will hit the target at an appropriate (right?) angle. Remember from our study of cyclotrons (Chapter 7) that magnetic fields will cause charged particles to move in a circular path. If we choose our magnet carefully (field strength) we can get the electron beam to go through a 90° *or* a 270° turn, sending them toward the target.

Unfortunately, the electrons are not all traveling at exactly the same velocity—they have slightly different energies. When they hit the 90° magnetic field, those that are traveling a little faster than the rest ($E+$) are bent through an angle slightly larger than 90°. Those that are traveling a little slower ($E-$) are bent through an angle that is somewhat lower than 90°. This causes the electron beam to broaden a bit, with lower-energy electrons at one end and higher-energy electrons at the other (fig. 13.15). The effect is similar to that of a prism, which spreads out visible light according to energy. For this reason, 90° bending magnets are called **chromatic**.

The same problem is encountered in 270° bending magnets; the electrons are still spread out according to energy. However, running them through 270° causes them to converge at a certain point after exiting the magnet (fig. 13.16). If left alone, the electrons would again diverge after this point. This point is an excellent place for

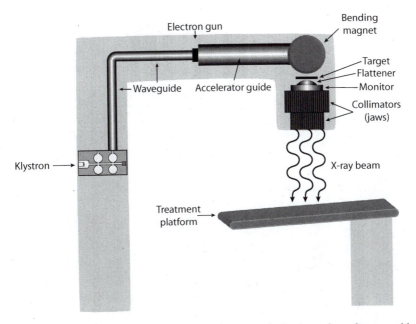

FIGURE 13.14   A MV X-ray machine with a klystron and a horizontal accelerator guide.

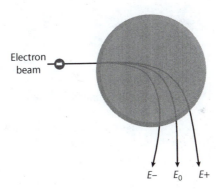

FIGURE 13.15   Energy dispersal effect of a 90° bending magnet.

the target. By doing this, we can obtain a very small focal spot which results in a narrow X-ray beam and minimal penumbra. Because of this, 270° bending magnets are called **achromatic**. As you may have already figured out, the focal spot on chromatic magnet machines is larger.

Achromatic magnets are more commonly used in medical linear accelerators, and since such a narrow beam is rarely needed, it needs to be broadened. This is done by the **flattener** (or flattening filter), which is sandwiched between the target and the monitor in figures 13.13 and 13.14. Another reason to use a flattening filter is that the photon beam tends to be rather intense in the center, and drops off rapidly from that point. For therapeutic applications it is better to have a more uniform beam. Flattening filters accomplish both tasks. Because they modify photon beams

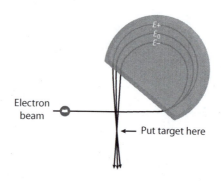

FIGURE 13.16   Energy convergence effect of a 270° bending magnet.

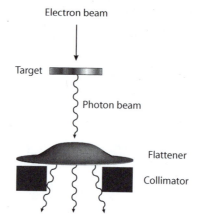

FIGURE 13.17   The dispersal effect of a flattening filter on an X-ray beam.

(low probability of interacting with matter), flatteners are chunks of metal that have different shapes depending on the energy of the photon beam—generally speaking, they are thicker in the center where the photon beam is most intense (fig. 13.17).

After going through the flattener, the resulting X-ray beam should be flat and symmetric. They are both measures of the uniformity of the beam. **Flatness** is measured by looking at the variations in the intensity over the center 80% of the beam. Intensity should not vary by more than ±3% across this region (fig. 13.18a). **Symmetry** is determined by looking at the overall intensity between the two halves of the beam—it should not vary by more than 2% (fig. 13.18b). If the flattener is not properly aligned to the beam, an asymmetric beam will result.

Finally, we should mention **safety interlocks**—an important component of any radiation-generating device. Safety interlocks are switches that will shut down the X-ray beam or the entire machine if a particular safety mechanism is not in place. For example, there is a switch on the door to the treatment room to ensure it is closed before exposing the patient. There are also switches to detect proper placement of filters and collimators, as well as functioning warning lights. Instrument software will also have safety interlocks checking flatness, symmetry, and patient dose.

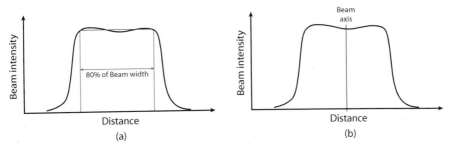

FIGURE 13.18   Measurement of (a) beam flatness and (b) beam symmetry.

Electron beams can be generated by the same machines discussed here—just remove the target so the electron beams hit the patient instead of X-rays. Electron beams will also need to be broadened. As they emerge from the accelerator, these beams are only 1–2 mm in diameter. A thin metal foil will generally do the trick (remember Rutherford's gold foil scattering experiment?). These are called **scattering foils**. They are thin because electrons have a relatively high probability of interacting with matter and thicker chunks of metal would absorb all of the electrons. Additionally, an **electron applicator** or **cone** is used to help collimate the beam beyond the jaws. Electrons are more likely to be scattered by the air they travel through, so these cones will need to be fairly close to the patient.

QUESTIONS

1. Increasing the current on a conventional tube increases the X-ray intensity. If greater intensity is needed, why can't the current be cranked up as needed?

2. On a single plot, overlay the three X-ray spectra produced by the same conventional X-ray tube set at three different voltages.

3. Tungsten, molybdenum, and copper are all are used for X-ray production. For each anode material, calculate the electron energy needed for 100% efficiency of photon production. For the highest value, also determine the electron velocity. Does this seem reasonable for a clinical setting?

4. Give two reasons why a ^{60}Co irradiator would be preferred for radiation therapy over a 500 kV orthovoltage machine.

5. Explain the general trend of greater PDD with X-ray energy observed in figure 13.6 for the five curves. Why do the highest-energy curves peak between 0 and 5 cm depth?

6. Give three reasons why megavoltage machines are favored for radiation therapy over orthovoltage units. Briefly describe each reason.

7. Carefully make drawings like figure 13.7b and 13.7c, except change only the SDD. Which gives a larger penumbra? Repeat, this time moving the patient a little closer (reduce SSD) to the source. Which of these latter two drawings shows a smaller penumbra? You might want to use a computer-drawing program or graph paper to insure consistency between your drawings.

8. Briefly explain the main similarity and difference between the Bragg peak associated with charged particles' interactions with matter and the fact that maximum dose is delivered inside a patient by high-energy photons.

9. Briefly explain how radiation therapy patients can be exposed to electron radiation during a photon treatment by a 4 MV machine, that is, the target and flattener are in place and SSD = 160 cm. Calculate the minimum possible LET (in air) of an electron produced by a 4 MV machine.

10. Give two unrelated reasons why the electrons produced in the previous problem are very likely to have a higher LET.

11. What is the purpose of a magnetron? A klystron?

12. Sketch a diagram of a bent-beam accelerator. Label all major components.

13. Inorganic scintillator and semiconductor detectors are much more efficient at detecting high-energy photons. Why aren't they used in place of the ionization chamber in megavoltage machines?

14. What is an X-ray flattening filter? How does it differ from a scattering foil?

15. Chromatic also has meaning in music. Is there an analogy between this definition and the bending magnets described in this chapter? Explain your answer.

Bibliography

Baum, Edward M., Harold D. Knox, and Thomas R. Miller. 2002. *Nuclides and isotopes: Chart of the nuclides*. 16th ed. Bethesda, MD: Lockheed Martin.

Bertulani, Carlos A., and Helio Schechter. 2002. *Introduction to nuclear physics*. New York: Nova Science.

Chase, Grafton D., and Joseph L. Rabinowitz. 1967. *Principles of radioisotope methodology*. 3rd ed. Minneapolis: Burgess.

Cherry, Simon R., James A. Sorenson, and Michael E. Phelps. 2003. *Physics in nuclear medicine*. 3rd ed. Philadelphia: Saunders.

Choppin, Gregory, Jan Rydberg, and Jan-Olov Liljenzin. 2001. *Radiochemistry and nuclear chemistry*. 3rd ed. Oxford: Butterworth-Heinemann.

Dunlap, Richard A. 2004. *The physics of nuclei and particles*. Belmont, CA: Brooks/Cole.

Ehmann, William D., and Diane E. Vance. 1991. *Radiochemistry and nuclear methods of analysis*. New York: Wiley.

Friedlander, Gerhart, Joseph W. Kennedy, and Julian M. Miller. 1981. *Nuclear and radiochemistry*. 3rd ed. New York: Wiley.

Hála, Jiří, and James D. Navratil. 2003. *Radioactivity, ionizing radiation and nuclear energy*. 2nd ed. Berkova, Czech Republic: Konvoj.

Heath, R. L. 1997. *Scintillation spectrometry: Gamma-ray spectrum catalogue*. vol. 1. 2nd ed. Idaho National Laboratory, http://www.inl.gov/gammaray/.

Hendee, William R., Geoffrey S. Ibbott, and Eric G. Hendee. 2005. *Radiation therapy physics*. 3rd ed. New York: Wiley.

Khan, Faiz M. 1994. *The physics of radiation therapy*. 2nd ed. Baltimore: Williams & Wilkins.

Knoll, Glenn F. 2000. *Radiation detection and measurement*. 3rd ed. New York: Wiley.

Lawrence Berkeley Laboratory. 2007. The ABC's of Nuclear Science, http://www.lbl.gov/abc/.

Lieser, Karl H. 2001. *Nuclear and radiochemistry: Fundamentals and applications*. 2nd ed. Berlin: Wiley-VCH.

Loveland, Walter, David J. Morrissey, and Glenn T. Seaborg. 2006. *Modern nuclear chemistry*. New York: Wiley.

Saha, Gopal B. 1998. *Fundamentals of nuclear pharmacy*. 4th ed. New York: Springer.

Semat, Henry and John R. Albright. 1972. *Introduction to atomic and nuclear physics*. 5th ed. New York: Holt, Rinehart and Winston.

Settle, Frank A. 2007. Alsos Digital Library for Nuclear Issues. Washington and Lee University, http://alsos.wlu.edu.

Stanton, Robert, and Donna Stinson. 1996. *Applied physics for radiation oncology*. Madison, WI: Medical Physics.

Walker, Samuel J. 2004. *Three mile island: A nuclear crisis in historical perspective*. Berkeley: University of California Press.

Wolfson, Richard. 1991. *Nuclear choices: A citizen's guide to nuclear technology*. Cambridge, Mass: MIT Press.

Appendix A: Useful Constants, Conversion Factors, SI Prefixes, and Formulas

CONSTANTS

Speed of light	$c = 2.998 \times 10^8$ m/s
Planck's constant	$h = 6.626 \times 10^{-34}$ J·s
Avogadro's number	6.022×10^{23} things/mol
W-quantity in air	33.85 eV/IP
Proton mass	1.673×10^{-27} kg $= 1.007276$ u
Neutron mass	1.675×10^{-27} kg $= 1.008665$ u
Electron mass	9.109×10^{-31} kg $= 5.486 \times 10^{-4}$ u
Boltzmann's constant	1.3807×10^{-23} J/K $= 8.63 \times 10^{-5}$ eV/K

CONVERSION FACTORS

$1\ u = 1.66054 \times 10^{-24}$ g $1\ Ci = 3.7 \times 10^{10}$ dps $1\ \mu Ci = 2.22 \times 10^6$ dpm

$1\ Bq = 1$ dps 5.256×10^5 min $= 1$ a $1\ J = 6.242 \times 10^{18}$ eV

$1\ J = 1\ kg \cdot m^2 / s^2$ $1\ R = \dfrac{2.58 \times 10^{-4}\ C}{kg}$ $1\ J/kg = 1\ Gy$ 931.5 MeV $= 1$ u

$1\ rad = \dfrac{100\ erg}{g} = \dfrac{10^{-2}\ J}{kg} = 1\ cGy$ 365.2 d $= 1$ a   1 rem $= 10^{-2}$ Sv $= 1$ cSv $= 10$ mSv

SI PREFIXES

Prefix	Symbol	Factor
tera	T	10^{12}
giga	G	10^9
mega	M	10^6
kilo	k	10^3
centi	c	10^{-2}
milli	m	10^{-3}
micro	μ	10^{-6}
nano	n	10^{-9}
pico	p	10^{-12}
femto	f	10^{-15}

FORMULAS

$$c = \lambda \nu \qquad E = h\nu \qquad A = Z + N$$

$$\ln \frac{A_2}{A_1} = -kt \qquad A_2 = A_1 e^{-kt} \qquad k = \frac{\ln 2}{t_{1/2}}$$

$$\ln \frac{N_2}{N_1} = -kt \qquad N_2 = N_1 e^{-kt} \qquad A = kN$$

$$\% \text{ efficiency} = \frac{\text{cpm}}{\text{dpm}} \times 100\% \qquad \text{specific activity} = \frac{\text{activity}}{\text{mass}}$$

Secular equilibrium: $N_B = \dfrac{k_A}{k_B} N_A (1 - e^{-k_B t}) \qquad A_B = A_A (1 - e^{-k_B t})$

at equilibrium: $A_B = A_A \dfrac{N_B}{N_A} = \dfrac{k_A}{k_B} = \dfrac{t_{1/2(B)}}{t_{1/2(A)}} \quad$ or $\quad N_A k_A = N_B k_B$

Transient equilibrium: $\dfrac{N_B}{N_A} = \dfrac{t_{1/2(B)}}{t_{1/2(A)} - t_{1/2(B)}} \qquad \dfrac{A_A}{A_B} = 1 - \dfrac{t_{1/2(B)}}{t_{1/2(A)}}$

$$t_{max} = \left| \frac{1.44 t_{1/2(A)} t_{1/2(B)}}{\left(t_{1/2(A)} - t_{1/2(B)} \right)} \right| \times \ln \frac{t_{1/2(A)}}{t_{1/2(B)}}$$

$$\sigma = \sqrt{\bar{x}} \qquad RSD = \sigma / \bar{x} s \qquad \sigma_{mean} = \sigma / \sqrt{N} = \sigma / \sqrt{t}$$

$$\chi^2 = \frac{1}{\bar{x}} \sum (\bar{x} - x)^2 \qquad E = mc^2$$

$$KE = \tfrac{1}{2} mv^2 \qquad KE = \left(\frac{1}{\sqrt{1 - (v^2/c^2)}} - 1 \right) mc^2 \qquad v = c \sqrt{1 - \left(\frac{mc^2}{KE + mc^2} \right)^2}$$

$$LET = SI \times W \qquad \text{Range} = R = \frac{E}{LET} \qquad RSP = \frac{R_{air}}{R_{abs}}$$

$$SI = \frac{4500 \text{ IP/m}}{v^2/c^2} \qquad E_{SC} = \frac{E_0}{1 + \left(E_0/0.511 \right) \times (1 - \cos \theta)}$$

$$I = I_0 e^{-\mu x} \qquad \ln\frac{I}{I_0} = -\mu x \qquad HVL = \frac{\ln 2}{\mu}$$

$$TVL = \frac{\ln 10}{\mu} \qquad \mu_m = \frac{\mu}{\rho} \qquad \mu_a = \frac{\mu \times \text{atomic mass}}{\rho \times \text{Avogadro's number}}$$

$$\mu_{eff} = \frac{\ln(I_0/I)}{x} \qquad \mu_{eff} = \frac{\ln 2}{HVL}$$

$$E_{CE} = \frac{E_0^2}{E_0 + 0.2555} \qquad E_{BS} = \frac{E_0}{1 + (3.91 \times E_0)} \qquad E_0 = E_{CE} + E_{BS}$$

$$E_{tr} = -Q \times \left(\frac{A_A + A_X}{A_A}\right) \quad E_{ecb} \approx 1.11 \times \left(\frac{A_A + A_X}{A_A}\right) \times \left(\frac{Z_A Z_X}{A_A^{1/3} + A_X^{1/3}}\right) \quad E_{KC} \approx E_{KX} \times \left(\frac{A_X}{A_C}\right)$$

$$N_B = N_A \Phi \sigma t \qquad N_B = \frac{\sigma \Phi N_A}{k}\left(1 - e^{-kt}\right) \qquad A_0 = \sigma \Phi N_A \left(1 - e^{-kt}\right)$$

$$A_{sat} = \sigma \Phi N_A \qquad A_0 = \sigma \Phi N_A kt$$

$$K_T = 0.80 \times \frac{Z_1 Z_2}{\sqrt[3]{A_1} + \sqrt[3]{A_2}} \qquad K_1 = K_T\left(\frac{A_2}{A_1 + A_2}\right) \qquad K_2 = K_T\left(\frac{A_1}{A_1 + A_2}\right)$$

$$\frac{\text{rate}_A}{\text{rate}_B} = \sqrt{\frac{MW_B}{MW_A}} \qquad H = \text{dose} \times Q \qquad f = \frac{\text{dose in medium}}{\text{dose in air}}$$

$$X = \frac{\Gamma \cdot A \cdot t}{d^2} \qquad X = \frac{\Gamma \cdot A \cdot t}{d^2} B \qquad B = I/I_0$$

$$f = 3.5 \times 10^{-4} ZE$$

Appendix B: Periodic Table of the Elements

1 **H** 1.008																	2 **He** 4.003
3 **Li** 6.941	4 **Be** 9.012											5 **B** 10.81	6 **C** 12.01	7 **N** 14.01	8 **O** 16.00	9 **F** 19.00	10 **Ne** 20.18
11 **Na** 22.99	12 **Mg** 34.31											13 **Al** 26.98	14 **Si** 28.08	15 **P** 30.97	16 **S** 32.06	17 **Cl** 35.45	18 **Ar** 39.95
19 **K** 39.10	20 **Ca** 40.08	21 **Sc** 44.96	22 **Ti** 47.87	23 **V** 50.94	24 **Cr** 52.00	25 **Mn** 54.94	26 **Fe** 55.84	27 **Co** 58.93	28 **Ni** 58.69	29 **Cu** 63.55	30 **Zn** 65.41	31 **Ga** 69.72	32 **Ge** 72.64	33 **As** 74.92	34 **Se** 78.96	35 **Br** 79.90	36 **Kr** 83.8
37 **Rb** 85.47	38 **Sr** 87.62	39 **Y** 88.90	40 **Zr** 91.22	41 **Nb** 92.91	42 **Mo** 95.94	43 **Tc** (99)	44 **Ru** 101.1	45 **Rh** 102.9	46 **Pd** 106.4	47 **Ag** 107.9	48 **Cd** 112.4	49 **In** 114.8	50 **Sn** 118.7	51 **Sb** 121.8	52 **Te** 127.6	53 **I** 126.9	54 **Xe** 131.3
55 **Cs** 132.9	56 **Ba** 137.3	*	72 **Hf** 178.5	73 **Ta** 180.9	74 **W** 183.8	75 **Re** 186.2	76 **Os** 190.2	77 **Ir** 192.2	78 **Pt** 195.1	79 **Au** 197.0	80 **Hg** 200.6	81 **Tl** 204.4	82 **Pb** 207.2	83 **Bi** 209.0	84 **Po** (209)	85 **At** (210)	86 **Rn** (222)
87 **Fr** (223)	88 **Ra** (226)	**	104 **Rf** (257)	105 **Db** (262)	106 **Sg** (266)	107 **Bh** (267)	108 **Hs** (269)	109 **Mt** (268)	110 **Ds** (271)	111 **Rg** (272)							

*	57 **La** 138.9	58 **Ce** 140.1	59 **Pr** 140.9	60 **Nd** 144.2	61 **Pm** (145)	62 **Sm** 150.4	63 **Eu** 152.0	64 **Gd** 157.2	65 **Tb** 158.9	66 **Dy** 162.5	67 **Ho** 164.9	68 **Er** 167.2	69 **Tm** 168.9	70 **Yb** 173.0	71 **Lu** 175.0
**	89 **Ac** (227)	90 **Th** 232.0	91 **Pa** 231.0	92 **U** 238.0	93 **Np** (237)	94 **Pu** (244)	95 **Am** (243)	96 **Cm** (247)	97 **Bk** (247)	98 **Cf** (251)	99 **Es** (252)	100 **Fm** (257)	101 **Md** (258)	102 **No** (259)	103 **Lr** (261)

Appendix C:
Table of Chi-Squared

Degrees of freedom	Probability of observing an outcome at least as large as the chi-squared value						
	0.99	0.95	0.90	0.50	0.10	0.05	0.01
2	0.020	0.103	0.211	1.386	4.605	5.991	9.210
3	0.115	0.352	0.584	2.366	6.251	7.815	11.35
4	0.297	0.711	1.064	3.357	7.779	9.488	13.28
5	0.554	1.145	1.610	4.351	9.236	11.07	15.09
6	0.872	1.635	2.204	5.348	10.65	12.59	16.81
7	1.239	2.167	2.833	6.346	12.02	14.07	18.48
8	1.646	2.733	3.490	7.344	13.36	15.51	20.09
9	2.088	3.325	4.168	8.343	14.68	16.92	21.67
10	2.558	3.940	4.865	9.342	15.99	18.31	23.21
11	3.053	4.575	5.578	10.34	17.28	19.68	24.73
12	3.571	5.226	6.304	11.34	18.55	21.03	26.22
13	4.107	5.892	7.042	12.34	19.81	22.36	27.69
14	4.660	6.571	7.790	13.34	21.06	23.69	29.14
15	5.229	7.261	8.547	14.34	22.31	25.00	30.58
16	5.812	7.962	9.312	15.34	23.54	26.30	32.00
17	6.408	8.672	10.09	16.34	24.77	27.59	33.41
18	7.015	9.390	10.87	17.34	25.99	28.87	34.81
19	7.633	10.12	11.65	18.34	27.20	30.14	36.19
20	8.260	10.85	12.44	19.34	28.41	31.41	37.57
21	8.897	11.59	13.24	20.34	29.62	32.67	38.93
22	9.542	12.34	14.04	21.34	30.81	33.92	40.29
23	10.20	13.09	14.85	22.34	32.01	35.17	41.64
24	10.86	13.85	15.66	23.34	33.20	36.42	42.98
25	11.52	14.61	16.47	24.34	34.38	37.65	44.31
26	12.20	15.38	17.29	25.34	35.56	38.89	45.64
27	12.88	16.15	18.11	26.34	36.74	40.11	46.96
28	13.57	16.93	18.94	27.34	37.92	41.34	48.28
29	14.26	17.71	19.77	28.34	39.09	42.56	49.59
30	14.95	18.49	20.60	29.34	40.26	43.77	50.89
40	22.16	26.51	29.05	39.34	51.81	55.76	63.69
50	29.71	34.76	37.69	49.33	63.17	67.51	76.15

60	37.49	43.19	46.46	59.33	74.40	79.08	88.38
70	45.44	51.74	55.33	69.33	85.53	90.53	100.4
80	53.54	60.39	64.28	79.33	96.58	101.88	112.3
90	61.75	69.13	73.29	89.33	107.6	113.1	124.1
100	70.07	77.93	82.36	99.33	118.5	124.3	135.8

Index